Decolonising Political Concepts

This book presents a transdisciplinary and transnational challenge to the enduring coloniality of political concepts, discussing the need to decolonise both their theoretical constructions as well as their substantive translations into practices.

Despite the acclaimed twentieth-century decolonisation waves, coloniality still remains in subtle and obvious practices, in visible and invisible mechanisms of power, and in the privileging of certain knowledges and the dismissing of others. *Decolonising Political Concepts* critically addresses the role political concepts play in the continuing legacies of colonialism and ongoing coloniality. This book, building on postcolonial and decolonial thinkers and ideas, demonstrates how concepts may be used as oppressing political and epistemological tools. By presenting efforts to decolonise political concepts, the book signals the potential for genuinely postcolonial academic and political contexts. Bringing together scholars from different disciplines and engaging with a wide array of geographical contexts, the chapters examine concepts such as agency, violence, freedom, or sovereignty. This book enables readers to critically engage with concepts used in political discourse and allows them to reflect on their impact and alternatives.

It will appeal to graduate students and scholars from international relations, social sciences, or philosophy, as well as to socio-political actors engaged in decolonisation agendas.

Valentin Clavé-Mercier is currently a Postdoctoral Researcher at the Complutense Institute for International Studies (ICEI) at the Universidad Complutense de Madrid (Spain). His overall research interest lies in how non-Western and decolonial political ontologies and praxis contribute to the rearticulation of contemporary political thought and political imaginaries. His most recent research focuses on discourses and practices of Indigenous sovereignty, more specifically on their deployment by Māori in Aotearoa/New Zealand. His areas of interest include decolonial/postcolonial studies, Indigenous politics, contentious politics, sovereignty studies, political geography, and identity politics. He is the author of "Politics of Sovereignty: Settler Resonance and Māori Resistance in Aotearoa/New Zealand" (2022).

Marie Wuth is a Postdoctoral Researcher at the University of Hamburg. She is a political and moral philosopher specialising in social and political ontology, environmental ethics, democratic theory, early modern philosophy, French theory, and decolonial and feminist theory. Her research focuses on the question of the political, agency, identity, and the power of affects and images in politics. Additionally, she is interested in the impact of relations of power and societal structures for conceptualisations and the relation of nature and politics. Her most recent publications include "Hate. Imaginary Roots and Fatal Dynamics of a Complex Relations" (2022) and "Circular Politics. Potentials, Limits and Boundaries of an Arendtian Nature-Politics" (2021).

Routledge Research on Decoloniality and New Postcolonialisms

Series Editor

Mark Jackson

Senior Lecturer in Postcolonial Geographies, School of Geographical Sciences, University of Bristol, UK.

Routledge Research on Decoloniality and New Postcolonialisms is a forum for original, critical research into the histories, legacies, and life-worlds of modern colonialism, postcolonialism, and contemporary coloniality. It analyses efforts to decolonise dominant and damaging forms of thinking and practice, and identifies, from around the world, diverse perspectives that encourage living and flourishing differently. Once the purview of a postcolonial studies informed by the cultural turn's important focus on identity, language, text and representation, today's resurgent critiques of coloniality are also increasingly informed, across the humanities and social sciences, by a host of new influences and continuing insights for different futures: indigeneity, critical race theory, relational ecologies, critical semiotics, posthumanisms, ontology, affect, feminist standpoints, creative methodologies, post-development, critical pedagogies, intercultural activisms, place-based knowledges, and much else. The series welcomes a range of contributions from socially engaged intellectuals, theoretical scholars, empirical analysts, and critical practitioners whose work attends, and commits, to newly rigorous analyses of alternative proposals for understanding life and living well on our increasingly damaged earth.

This series is aimed at upper-level undergraduates, research students and academics, appealing to scholars from a range of academic fields including human geography, sociology, politics and broader interdisciplinary fields of social sciences, arts and humanities.

The Coloniality of Modern Taste
A Critique of Gastronomic Thought
Zilkia Janer

Coloniality and Decolonization in the Nordic Region
Edited by Adrián Groglopo and Julia Suárez-Krabbe

African Epistemologies in Higher Education Research
Kolawole Samuel Adeyemo

Decolonising Political Concepts
Edited by Valentin Clavé-Mercier and Marie Wuth

For more information about this series, please visit: https://www.routledge.com/Routledge-Research-on-Decoloniality-and-New-Postcolonialisms/book-series/RRNP

Decolonising Political Concepts

Edited by Valentin Clavé-Mercier and Marie Wuth

Routledge
Taylor & Francis Group
LONDON AND NEW YORK

First published 2024
by Routledge
4 Park Square, Milton Park, Abingdon, Oxon OX14 4RN

and by Routledge
605 Third Avenue, New York, NY 10158

Routledge is an imprint of the Taylor & Francis Group, an informa business

Open Access is funded by the European Union Horizon 2020 Research & Innovation programme under the Marie Skłodowska-Curie grant agreement No. 754326.

British Library Cataloguing-in-Publication Data
A catalogue record for this book is available from the British Library

ISBN: 978-1-032-27591-8 (hbk)
ISBN: 978-1-032-27597-0 (pbk)
ISBN: 978-1-003-29346-0 (ebk)

DOI: 10.4324/9781003293460

Typeset in Times New Roman
by SPi Technologies India Pvt Ltd (Straive)

This book is dedicated to those who are thinking otherwise, and to those who want to.

Contents

Contributors

Karim Barakat is currently a Postdoctoral Fellow at the Center for Arts and Humanities at the American University of Beirut. His research focuses on what insight descriptive accounts, whether historical, metaphysical, or epistemological, can offer for social and political critique. He is also developing an account of political scepticism, which contributes to minimizing antagonism in political disagreement. Moreover, his research also examines the use of history in Arabic Political Philosophy. He has published several articles on the role of descriptive accounts in political theorizing such as "Understanding Power Through Governmentality" (2020) and "Grounding Reasonableness in Rawls" (2020).

Radhika Chitkara is an Assistant Professor at National Law School, Bangalore, India. Her work has been embedded within the civil liberties, women's, and land rights movements in northern and eastern India. Her research interests include issues of criminal justice, civil liberties, human rights, and feminist legal theories. She is currently pursuing a PhD in "Special Procedures under Anti-Terror Laws and the Right to Fair Trial: An Empirical and Doctrinal Study of Investigations and Prosecutions under the UAPA" from NLSIU.

Cecilia Cienfuegos Martínez is a PhD Student at Carlos III University. Her work focuses on sexual violence, political philosophy, and epistemology. From the intersection of feminist theory, decolonial theory, and testimonial accounts of sexual violence, her research tries to construct an alternative epistemology of this form of violence. She is the author, together with Ana Abril, of "Cannibalising Violence: Rethinking the Cannibal in Order to Theorise an Unthinkability of Sexual Violence" (2021), in Champion, G. (Ed.) *Interdisciplinary Essays on Cannibalism: Bites Here and There*.

Valentin Clavé-Mercier is currently a Postdoctoral Researcher at the Complutense Institute for International Studies (ICEI) at the Universidad Complutense de Madrid (Spain). His overall research interest lies in how non-Western and decolonial political ontologies and praxis contribute to the rearticulation of contemporary political thought and political imaginaries.

His most recent research focuses on discourses and practices of Indigenous sovereignty, more specifically on their deployment by Māori in Aotearoa/New Zealand. His areas of interest include decolonial/postcolonial studies, Indigenous politics, contentious politics, sovereignty studies, political geography, and identity politics. He is the author of "Politics of Sovereignty: Settler Resonance and Māori Resistance in Aotearoa/New Zealand" (2022).

Laura Galián is Assistant Professor in the Department of Arab and Islamic Studies and Oriental Studies, Universidad Autónoma de Madrid (Spain). Before, she was a postdoctoral researcher in the European H2020 project RETOPEA: Religious Toleration and Peace at Universidad de Granada and a Juan de la Cierva research fellow at the same university. Her research interests and publications deal with the history of concepts, the history of ideas, and the Middle East. She has recently published the book *Colonialism, Transnationalism and Anarchism in the South of the Mediterranean* (2020).

Oscar Guardiola-Rivera is Professor of Political Philosophy & Human Rights at Birkbeck College, University of London. His research and teaching research and teaching focus on peace-making, mediation and conflict-resolution, political philosophy and legal theory, art & human rights, human rights and international law, peoples' tribunals, advanced constitutional law and ethics, with an emphasis on decolonial and visual methodologies, political economy, and racial justice. He is the award-winning author of *What If Latin America Ruled the World* (2010), *Story of a Death Foretold* (2013), and the poetic novella *Night of the World* (2020). He currently works on a memoir ethnography titled *Journal of the Twenty-First Century Years of Plague*.

Henrike Kohpeiß is a Philosopher and a Postdoctoral Researcher at the CRC Affective Societies at Freie Universität Berlin. She works in the fields of critical theory, Black Studies, and affect theory and is interested in critiques of the subject, histories of violence, and questions of coloniality and race. She is the author of *Bürgerliche Kälte. Affekt und koloniale Subjektivität* (2023).

Shahin Nasiri is a Lecturer in philosophy at Tilburg University. His areas of interest include political theory, theories of freedom, phenomenology, critical theory, migration and citizenship, and genealogy of resistance movements. He is the author of "Investigating the 1981 Massacre: On the Law-Constituting Force of Violence" (2022), "Freedom and the Imaginary Dimension of Society" (2022), and *The Repressed Voices of the Iranian Revolution* (2020).

Laurencia Sáenz Benavides is a Philosophy Lecturer at the School of Philosophy of the University of Costa Rica. She works on moral and political philosophy, feminist philosophy, critical philosophy of race, and philosophy of emotions. She is the author of the book *Instrumentos de poder: el rol de las emociones en la opresión* (2021).

Astha Saxena is a Law Researcher whose work lies in the rights to property and land, laws of land reform, acquisition, and studying the implementation of the 2006 Forest Rights Act. She also works with various groups of lawyers and social researchers to expand the boundaries of legal education. In her doctoral studies, she is exploring the relevance of legislative paradigms towards the realisation of socio-economic rights. She has authored *Land Law in India* (2020) and co-authored "Criminalisation of Adivasis and the Indian Legal System" (2022).

Rafael Verbuyst is a Historian and Anthropologist. He is currently a Postdoctoral Researcher at Ghent University's History Department – fully funded by a grant from the Flemish Scientific Fund. He is also a visiting research fellow at the University of Cape Town's Centre for African Studies. His research centres on the revival of indigenous identity in post-apartheid South Africa, land (re)claims, settler colonialism, decolonisation and resistance, and ethnographic methodology. He is the author of *Khoisan Consciousness: An Ethnography of Emic Histories and Indigenous Revivalism in Post-Apartheid Cape Town* (2022).

Ritu Vij is a Senior Lecturer in International Relations at the University of Aberdeen, UK. Her current work examines the limits of contemporary thinking on sovereignty, precarity, and subalternity in two ongoing collaborative projects: "Precarity: Poetic and Aesthetic Explorations", and "Political Imaginaries of the "Slum". Her recent work includes a co-edited collection, *Precarity and International Relations* (2020), "Precarity and the International", a special issue of *Globalizations*, articles on affective and housing precarity, human security and difference, and the cinema of precarity. Her earlier books are *Japanese Modernity and Welfare* (2007) and *Globalization and Welfare: A Critical Reader* (2007).

Marie Wuth is a Postdoctoral Researcher at the University of Hamburg. She is a political and moral philosopher specialising in social and political ontology, environmental ethics, democratic theory, early modern philosophy, French theory, and decolonial and feminist theory. Her research focuses on the question of the political, agency, identity, and the power of affects and images in politics. Additionally, she is interested in the impact of relations of power and societal structures for conceptualisations and the relation of nature and politics. Her most recent publications include "Hate. Imaginary Roots and Fatal Dynamics of a Complex Relations" (2022) and "Circular Politics. Potentials, Limits and Boundaries of an Arendtian Nature-Politics" (2021).

Preface: We Shall Dance Better

Oscar Guardiola-Rivera

When we launched what is now known worldwide as "the decolonial turn" back in the mid-1990s, our mind was not set on producing yet another sociological type ("coloniality", for instance). Instead, we wanted to place a bomb. To explode the framework within which the social sciences and the humanitarian arts of law and literature had been developed since at least the dawn of modernity.

Perhaps, in this respect, it would be best to speak of the "arrested development" of the sciences, of humanitarianism, and of the republican arts, which for us meant, quite simply, the making of things public.

We knew that a different kind of humanism and republicanism had emerged in the wake of the encounter between "Christians" and the societies that worked out or retained a bio-cultural wisdom, especially in the Americas, which made them resist the idea of progress toward perfection or achievement followed by stasis that had become a dogmatic principle of politics and the mind after the importation of the techniques of perspective to the practical knowledges of siege architecture, speculative geography, and surveyance which informed the expansionist enterprise of European kingly powers.

We knew such humanism and republicanism had actively stood against the association between a way of seeing and arguing with demonstrative pretensions, in which the axiomatic and the arbitrary go hand in hand, best exemplified by the defensive attitude and mentality expressed in the justifications of conquest and empire in the Americas put forward by jurist-theologian Juan Ginés de Sepúlveda during the famous debates of the 1550s. And we also knew that a different association could be made between the critique of the imperialist project and a different image system that valued differently what happens to be the case in the world and argued about it in a fallibilist manner that can concede our world formulas are at best approximations and continue work at them by making them present and concrete in bio-cultural sites that involve the affordances of rivers, the vulnerability of bodies, and the experimental power of judgement and the imagination.

Because of this, we focused our efforts on the concrete practices of internal colonialism, discipline and control of time or time orientation, as well as self-colonisation as forms of "arrested development" that damned Amerindian,

Afro, and creolised societies to a static past and foreclosed other futures, thereby condemning the sciences and humanitarian or republican arts to remain within a framework in which dialectics are affixed to an absolute viewpoint or subject, while at the same time insisting on the active presence and ongoing-ness of the signs, images, and concrete symbolisations produced and reinvented by those societies.

In other words, instead of producing trademarks or tokens like "coloniality" of… this and that to market them among the rapidly changing fashions of contemporary political communications and academia, or assuming that "decolonial" waves in the nineteenth or twentieth century had resulted in sufficiently clean and simple colonised/coloniser or victim/perpetrator binaries, to use the now ubiquitous language of human rights, we focused on the percepts and concepts left behind by "traditional" (meaning demonstrative, arbitrary, and axiomatic) thought, historically associated with the expansionist enterprise, and the habits of sensing and familiar sense after its demise as philosophy (Castro-Gomez et al., 1999).

For us, the call for a frame of reference in which everything has its place after the fall of all ultimate points of reference, to seize a political philosophy, a concept, or something to hold onto, dissimulated under the pretext that such "centre" or "ground" would better guide us to the promised land, was but a repetition and reinvention, for the new speculative global markets and relations, of the older ways of seeing that organised the visual world as the universe was once thought to be arranged for God. Also, a mask for nothing more than aggression, the desire to appropriate and take hold of a philosophy, a landscape, and everything in it, in the way that the schools of old used to devour each other. That predatory scene was never for us a matter of abstract theory, since the apotheosis of war and plunder surrounded us in the lands of our childhood when we launched the modernity/coloniality research project in the 1990s, and such apotheosis has now viral, global reach.

The present book shares and takes further that spirit and focus on percepts and concepts, political concepts, as much as the critical (as opposed to "traditional") approach that animated the modernity/coloniality project in the first place. Not only it takes it up but runs with it! Its starting point is that the study of politics, law, and literature in the West largely stands on the premise that such things as politics, the communication of signs and symbolisation, laws, and kinds of value, as well as literature begin with letters and images posited and made literary or civil within a proper, ultimate frame of reference. Against the call for a frame of reference, emerging with the acceptance of analytic geometry and calculation not only as an approximation to the world but as a more real world hiding behind this one, which we could decrypt or decode if we were in possession of The Word or Cypher, the essays in this collection invite us to see better and see through. To shift our (speculative) geography and perspective. Not just to de-centre but to tell better stories about the so-called centre or centres both in the political philosophical or jurisprudential and geo-political sense.

To question the fantasy of harmony through calculation which has returned in full twenty-first-century garb, dressed up in shiny algorithmic or AI regalia that already contains quantification of what is spatially perceptible, abstraction according to conventional ways of seeing and thinking that may even be based on arbitrary axioms (so that the arbitrary and the axiomatic go hand in hand), and to translate our supposedly critical ideas and correct moral positions into parts of the imagination that are also points of contact with real relations – in motion, transiting, transformative, and transforming.

The first two parts of this book explore and question relations of power and knowledge embedded in the ways we see and conceptualise the world in which we dwell, often using the image of a fork on the road or an intersection. The third part of the book puts into practice other languages of refusal and resistance that may enable decolonial liberation across diverse geographical contexts and political movements.

Throughout, the emphasis is on movement or trans-motion. This entails a way of symbolizing, a symbol and a concept that is not a static one. Rather, it contains within itself a principle of differentiation. "Principle" in the powerful sense of a new beginning, and an idea capable of carrying that new beginning, in which all other pasts are present, into the future. As such, the very character of a concept is reworked in this book: it becomes a category that shows us the direction to follow. It is not only a matter of difference but one of difference and orientation.

However, and this is where the very idea of a decolonial turn, re-turn, or move makes better sense, the country in which a non-static symbol dwells, as in the Coyote or Donkey stories and word cinemas found throughout Amerindia, remains uncharted and not-yet subject to Euclidean mapping or geo-political lines of demarcation. If so, the political reason that emerges throughout the collaborations of this book is not the purer, more Euclidean reason that builds and elevates utopia to the status of a "shared value", belief, or faith. Would I dare saying that since such rationalist utopia is a power trip, monotheocratic, declared by executive decree or royal prerogative, maintained by strong willpower or constancy, premised on progress rather than process, and, therefore, Euclidean, European, and masculine, then this political reason is not utopian at all?

Or at least, this is not how utopia ought to look like. You will recall that the quality of stasis or static perfection (the "perfect" and "perfectible" communities of Second Scholastic political thought that still lie buried within our supposedly modern political and legal concepts) is actually an essential element, a quintessence or a deification (hence, our speaking before about the apotheosis of war and the market in the twenty-first century) and, therefore, also an element of the non-inhabitability of the Euclidean, white European, masculine utopia.

If so, to attain a more inhabitable time and place when faced with the apotheosis of war and market, with climate meltdown, and fascination of abomination, we must not only turn (no matter how "decolonial" that turn might seem) but also re-turn, go round, go inward, undercover and underground, like investigative detectives, like Forensic Architectures investigative detectives, or like surreal detectives such as PI Clem Snide.

The latter, you may recall, would try to solve a case by sitting back, listening at random to sound recordings made in a device not too dissimilar to today's Teenage Engineering PO portable mixers and sequencers "specially designed for cut-ins and overlays and you can switch from record to Playback without stopping" it. He would record the sounds of the forest outside the villa of the dead man, the rattle of dishes being washed, the sound of water and wind as he walks along a riverbed or the sea, as well as rave music to dance or even the toilet flushing. Later, he would randomly choose sections of the composite recording while watching TV so that he would listen only half- or sub-consciously.

To suggest that approach as an analogue for doing political and legal philosophy in the digital era might strike some as either too Kitsch or too Dada, at a time when the bourgeoisie has grown bored of both, declared them unfashionable, for they move fast, so fast they see no more than the surface glitter of a life too swift to be real. They/we are assailed by too many new things ever to find the depths, the roots in the survivance and ongoing poetics of the signs obscured by a cultural logic of the archive, the court, and the museum as a final resting place in which everything has its place. The rush of life past, that stormy wind, they/we call progress, though it is now too rapid for us to move with it.

But what if it is no longer the case that Amerindian naïve or Dada have subsided into history, and instead, before the facts of pandemics and climate meltdown bringing back the re-enchantment of nature, it is history that is subsiding into Dada and magical realism? And that can help when one knows not whether to go away or staying in the place where one's people dwell, when visited by war or with an infectious distemper (Taussig, 2021).

I want you to read this book because I know of no other work that takes for its central concern the political power of images, percepts, and concepts in a way that is neither utopian nor dystopian. But rather, a way that is no way, not on the road atlas, which is, after all, the only way to get to a place that is no place. The intellectual concern of this book is this matter of "westernisation", "progress", or a mimetic power so excessive and so fast we can no longer move with it. This is perhaps the central fact and concern of our times. If so, of course this book provides no ultimate answer, model, roadmap, or solution; it simply indexes and indicates, like a lighthouse that is not a wreckers' lantern, the way that cannot be gone. Its poetic image is, therefore, that of an enantiodromia. Like a donkey moving backwards to the forest, or a porcupine backing into a crevice. To better see and hear the challenges ahead and think of a more habitable place. It moves sideways or backwards looking forwards. It dances. That is a proper decolonial move.

References

Castro-Gomez, S., Guardiola-Rivera, O. & Millán de Benavides, C. (1999). Introducción: Poscolonialismo, o la crítica cultural del capitalismo tardío. In S. Castro-Gomez, O. Guardiola-Rivera & C. Millán de Benavides (Eds.), *Pensar (en) los intersticios. Teoría y práctica de la crítica poscolonial* (pp. 9–19). Bogotá: CEJA.

Taussig, M. (2021). Would a Shaman Help? *Critical Inquiry*, *47*(s2).

At the Crossroads of Coloniality, Power, and Knowledge

It Is Time to Decolonise Political Concepts

Valentin Clavé-Mercier and Marie Wuth

> *[C]oncepts have teeth, and teeth that bite through time.*
>
> (Simpson, 2014: 100)

The first two decades of the 21st century have seen renewed attention to how sustaining legacies of colonialism continue to reside within and to shape social relations, politics, human–nature interactions, culture, or education. In recent years, this critical inquiry crystallised in a myriad of decolonisation calls that have been raised in a variety of ways: mounting social pressure to confront systemic and institutionalised racism epitomised by movements such as Black Lives Matter; political processes aimed at breaking colonial bonds between polities such as the New Caledonian independence referendums or Barbados' rupture with the British monarchy; warnings regarding the challenges of the so-called Anthropocene and connected environmental injustices; the questioning of the role of museums in colonial processes culminating in campaigns such as the Museums Association's "Decolonising Museums"; or students-led movements to decolonise educational curriculums and universities such as the Rhodes Must Fall movement in South Africa, the "Why is my curriculum White?" campaign in the United Kingdom.[1]

Western contemporary thought has similarly received significant attention for its contributing role in colonisation processes and in sustaining what is known as 'coloniality' (Chatterjee, 1993; Ivison et al., 2000; Mignolo, 2021; Pagden, 1995; Said, 1993; Walker, 1987). More than maintaining a mere relation of complicity with colonialism, Western modern political thought has been exposed as being shaped via the experiences of colonial expansion and invasion, then in turn actively contributing to justify and underpin said phenomena (Anghie, 2005; Barker, 2005; Bonilla, 2015; Shaw, 2008). However, the particular role of political concepts in this process, as the tools through which this political thought is articulated, has rarely been properly dissected. Beyond a theoretical endeavour, examining the colonial roots and ramifications of some of the most used political concepts in public forums, in the media or in academia is directly linked to ongoing political concerns and struggles being deployed in the Global South and Global North. It is about confronting the

DOI: 10.4324/9781003293460-1

continuous reinforcement of limits and exclusions brought to life through the uncritical use and reproduction of certain political concepts in their Western modern understandings. Similarly, it is about the capacity of Western modernity's Others to express and articulate alternative and resistance.

This volume was born out of an interest to contribute to the decolonisation of modern political thought by foregrounding the different dimensions and articulations of coloniality contained in some of its central political concepts and in the theories and practices associated with them. Across *Decolonising Political Concepts*, researchers discuss and reflect upon specific political concepts against the background of postcolonial and decolonial theories and ideas. The contributors share the conviction that, in spite of the formal and alleged demise of colonial domination, coloniality – understood as an ontological order hierarchising modes of existence along a spectrum of humanity and, thus, validity – still endures. It endures in the relation between the dominant Western culture and 'other-ed' cultures, in the global economic arrangements, in social discrimination and oppression structures articulated through race or ethnic constructions, and in a wide array of other processes. Moreover, coloniality unfolds in the privileging of certain forms of knowledge, in the dismissing, ignoring, or silencing of others. Problematising and defying the entanglement of all these processes and their encapsulation into socio-political constructs are at the heart of this collection. Concretely, the contributions gathered here explore how predominant political concepts in Western political thought are beset with colonial remnants in their construction, formulation, and deployment. We contend that using them uncritically leads to naturalising a biased purview on political life that in turn shapes the available contemporary political thinking and praxis. In addition, this volume engages with concrete practices of colonised peoples and decolonial activists trying to reveal, work through, or subvert the colonial load and impact of political ideas surrounding them.

It is our firm belief that this collection represents a timely contribution in that it confronts one of the most important contemporary issues in the task of articulating just, inclusive, and emancipatory political thought: how to determine, imagine, and construct appropriate forms and spaces of political expression for the worldviews, thinking, claims, and aspirations of colonised peoples. Addressing this question needs to start with unpacking the entanglement of the predominant political thought with coloniality. Yet, it is an endeavour that cannot be circumscribed to critique. As necessary is to articulate ways to work through, re-think, or even overcome this predominant political thought and transform it into an (a)venue for postcolonial and decolonial struggles. It is this question, in its twofold character, that this volume delves into through a critical examination of some of the most common political concepts present in contemporary political thought. As already argued above, its originality lies precisely in its focus on political concepts and their relations to coloniality and decolonisation, an aspect often neglected in the existing literature.

We contend that political concepts are actually crucial to the work of decolonisation, thus requiring specific attention. Such significance stems from the fact that they provide the ground for theories and practices deployed in the political and public sphere and beyond. On the one hand, political concepts inform political communities' models, infrastructures, or juridical systems, as well as political movements, resistance practices, reforms, and political imaginaries. On the other, they shape our perception of and orientation in the world, as well as our interactions with other individuals, groups, or nations. Thus, political concepts are not neutral or innocuous but explicit tools of power and efficient vehicles for establishing or changing relations of domination. They convey and may impose certain perspectives, beliefs, values, and norms, being used and deployed as tools for social, political, and cultural control. As historical constructs, they are part of the colonial legacies that still permeate our contemporary world. The way academic and socio-political actors define and use them is largely mediated by traditions of political thought marked and framed by coloniality. Yet, at the same time, they may be articulated and put to work in ways that may trouble it.

Despite the increasing and far-reaching work of postcolonial and decolonial research, this aspect of political concepts is still too often silenced or ignored. This volume asserts the need to question how we understand the world that surrounds us, what concepts are used to produce meaning, and the impact these particular lenses have in and beyond the analytical process. In doing so it contributes and adds to recent literature on the decolonisation of epistemology (Bendix et al., 2020; Meghji, 2021; Menon, 2022; Shilliam, 2021; Wood, 2020). However, it departs from the more narrow and predominantly philosophical lenses adopted in these works to instead compile a transdisciplinary and transnational exploration of theoretical and practical ramifications of political concepts, oftentimes illustrated through contemporary socio-political issues and struggles. As such, the present collection intends to shed a light on what is still a blind spot in decolonial theory, while simultaneously introducing a decolonial perspective in multiple discourses and analyses across the social sciences and humanities. It is the hope of its editors and contributors that such a volume will offer useful reflections for students, scholars, and activists in their respective journeys to decolonise their discourse, practices, teaching, and thinking.

Decolonial Theory, Political Concepts, and the Ideological West

Postcolonial and decolonial thinkers and activists have spent decades unravelling the intellectual, political, and structural legacies of colonialism in our contemporary world. Although a strict distinction between decolonial and postcolonial approaches is not always expedient or even possible, and productive perspectives at times emerge from combining them,[2] this volume is chiefly situated and understandable within the context of decolonial theory. By "decolonial theory" we refer to a corpus of thinking that poses colonialism as a fundamental problem intrinsic to Western modernity and the colonised as a potential agent for radical epistemic, symbolic, and material change. Said corpus has been predominantly developed by Latin American, Caribbean, Chicanxs, and Afro-American thinkers in the latter part of the 20th and early 21st centuries, while claiming roots in the works of early 20th and mid-20th thinkers such as W.E.B Du Bois, Aimé Césaire, Maya Angelou, Frantz Fanon or Sylvia Wynter. In fact, decolonial thinkers themselves insist that decolonial thinking and practice have existed since the early stages of colonisation (Hartman, 1997; Maldonado-Torres, 2011). They also are often critical with the idea of decolonial 'theory' as a unified and homogeneous body, as well as with the artificial divide between theory and practice that is proper of Western modernity. Instead, decolonial theory as we understand it here is closely entangled with the actual political mobilisations and struggles of colonised and racialised people.

One of the central claims of this strand of critical theory is that we do not live in a postcolonial world – where 'postcolonial' would be understood as a profound transformation beyond a mere removal of explicit ruling by colonial powers. Such an argument led to the distinction between colonialism and decolonisation on the one hand, and coloniality and decoloniality on the other. The first two are commonly understood as defined and confined socio-historical processes, often located in the past. Colonialism is a particular political relation between peoples and polities in which one exercises territorial and juridical occupation over the other. On the contrary, decolonisation would be the official rupture of such a relation and the removal of formal structures of domination. Instead, coloniality and decoloniality refer to the underlying logics of these processes, to the mentalities and philosophies in which these processes are rooted. In his famous "Outline of Ten Theses on Coloniality and Decoloniality", Nelson Maldonado-Torres defines coloniality and decoloniality as "the logic, metaphysics, ontology, and matrix of power created by the massive processes of colonization and decolonization" (2016: 10).

Broadly speaking, it can be said that coloniality consists in a hierarchisation of modes of being, modes of knowing, modes of thinking, and modes of collective life along a civilised/savage dichotomy rooted in the colonial invasions in which European modernity stands as the benchmark for civilised status (Maldonado-Torres, 2018; Mignolo, 2000; Wynter, 1995). Across decolonial literature, coloniality is often dissected as a threefold concept. The coloniality of knowledge encapsulates the idea of an epistemic Eurocentrism according to which European modern rationality is construed as the only valid mode of knowledge production and is presented as universal (Grosfoguel, 2007; Quijano, 2007). The coloniality of power is described as a social, political, and economic hierarchisation of human beings following a racial classification in which whiteness is placed at the top (Quijano, 2000). As a consequence, race is converted into the determining factor for the positioning of subjects and groups in power relations and structures, thus leading to a Westernisation of life or an imitation of European models in all spheres of life in order to access power positions. Finally, the coloniality of being refers to the construction of colonised peoples as non-beings or "sub-humans" and foregrounds coloniality as a project of formatting colonised peoples into specific forms of being through the imposition of subjectivities defined by the colonial gaze and adapted to the colonial project (Maldonado-Torres, 2007; Wynter, 2003).

In its various facets, coloniality thus entails forms and processes of expropriation, domination, marginalisation, exclusion, and elimination (from symbolic to material). Of special significance for decolonial thinking is the fact that it can be sustained even once colonialism has formally ended. Indeed, coloniality stands for a particular ontological, epistemological, and political framework rooted in an alleged Euromodern superiority and produced through colonisation processes but exceeding them in time and space. Therefore, if in some aspects and contexts, the cultural, social, and political domination established during the European colonial period might have been formally and apparently defeated – notably through the acclaimed 20th-century 'decolonisation waves' – coloniality still endures in more subtle and complex forms such as visible and invisible mechanisms of power or the privileging of certain knowledges and the dismissing of others. "This leads to a situation of colonies where their native and colonized subjects continue experiencing vast forms of dispossession even after independence" (Maldonado-Torres, 2016: 17).

Alternatively, decoloniality encompasses the efforts to challenge and overcome such a framework and to undo the nexus of power, knowledge, and subjectivity that is coloniality. Decoloniality thus pursues the profound decolonisation of social, political, and cultural relations beyond the mere achievement of factual independence that is often associated with 'decolonisation'. It is therefore not only a project of critique but also a generative endeavour aiming at the actualisation of an alternative constitutively grounded in the recovery of plurality against the imposition of modernity/coloniality. As per Maldonado-Torres' seminal definition:

> decoloniality refers to efforts at rehumanizing the world, to breaking hierarchies of difference that dehumanize subjects and communities and that destroy nature, and to the production of counter-discourses, counter-knowledges, counter-creative acts, and counter-practices that seek to dismantle coloniality and to open up multiple other forms of being in the world.
>
> (2016: 10)

Such a genuine and profound decolonisation – in the sense of decoloniality – is thus an "unfinished project" (Maldonado-Torres, 2011). Unfinished because it is open-ended and because it is inherently plural and heterogeneous. Decoloniality has no definition of purity or perfection, nor an exact ideal to aspire to or a specific and defined plan of action to get there. It is an attitude based on a questioning of the existing coloniality and aimed at the generation of alternatives characterised by plurality and interconnections. "Diversality" (Bernabé et al., 1990), "transmodernity" (Dussel, 1995), or "pluriversality" (Escobar, 2020) are diverse articulations of decoloniality in action. They all point towards a decolonial horizon based on the recovery of the multiplicity of modes of existence and on the elimination of practices of hierarchisation and exclusion characteristic of coloniality.

In this brief overview of decolonial thought, one of the main theses of this critical enterprise must not go unmentioned. Namely, that coloniality is an integral part of European modernity rather than a by-product or an independent phenomenon. The emergence of many modern institutions in Europe depended on colonial processes, especially on colonial exploitation. Imaginaries, resources, and wealth extracted from the colonial context have been constitutive of the European modernity project. Ramón Grosfoguel (2013) further signals how the development of modernity – while an emancipatory project in Europe – was still marked by the dismissal or belittlement of colonised subjects' epistemologies and by the reproduction of colonial structures and imaginaries. Moreover, decolonial thinkers argue that European modernity, as a totalising project oriented to the construction of universals forged out of the European particularities, is inherently expansive, predatory, and colonial in order to actualise such totalisation and universality. In other words, the conditions of possibility of European modernity hinged on the expansion of colonialism, which simultaneously was a result of modern paradigms and worldviews – the main ones being a hierarchisation of human forms of being linked to European superiority claims and the construction of a civilisational mission (Dussel, 1993; Mignolo, 2000; Quijano, 1992; Wynter, 2003). Modernity and coloniality are thus constitutively entangled and, arguably, inseparable.

This entanglement means that many of the conceptual and analytical tools that were developed through European modernity, and that we still use today to access and understand reality, have colonial foundations and implications. An uncritical use of the central categories and concepts of modernity is thus often bound to a reproduction of coloniality. If we are to question and undo coloniality, we need to interrogate how we understand the world we live in, how we perceive it, through which analytical constructs, and what impact these have on our realities. Thought constructs are never left entirely to theory. They have practical foundations and consequences. How we articulate and deploy them may enable and/or prevent certain interventions in the world. The organisation of our socio-political subjectivities and lives is influenced by the concepts and underlying knowledge we use. As Bonaventura de Sousa Santos pointed out: "There is no knowledge that is not known by someone for some purpose. All forms of knowledge uphold practices and constitute subjects" (2014: 201). As such, the overarching question behind a possible decolonial rearticulation of political concepts, be it from academia or from movements politics, is what kind of practices and subjects do we want to create and foster through our use of language?

This is precisely why language has received significant attention in decolonial thinking. Indeed, recognising the entanglement of modernity and coloniality unavoidably poses the question of possible tensions and even contradictions between the task of decolonisation and the use of certain political categories in their inherited modern forms. Using the language of the coloniser – the language of European modernity – to express and understand the experience of the colonised and to articulate the decolonial seems problematic. As a consequence, Gloria Anzaldúa (2009) called women of colour to "speak in tongues", in a language of the subaltern different from the predominant white male language that would properly encapsulate their own experiences, culture, and thinking. In the same vein, Kwasi Wiredu (1995, 1996) advanced a recentring of African (and especially Akan) political philosophy as a theoretical repertoire better equipped to articulate practical resolutions to some of the continent's pressing challenges. Indeed, the language we use defines the contours of reality and possibility and thus circumscribes political imagination and praxis in potentially limiting ways. Audre Lorde's (2007) famous assertion that "the Master's tools will never dismantle the Master's house" comes to mind. Even when Western political thought and its categories are not necessarily the language of expression used by colonised peoples, the latter have often been historically forced to use them in order to be heard and/or understood.

However, several decolonial thinkers have argued for a politically emancipatory use of the coloniser's oppressive language. For instance, Frantz Fanon (1967) and Aimé Césaire (2001) defended the use of French as a means to create a new language capable of articulating and communicating the experiences of the colonised. New means of expression forged out of a colonially loaded point of departure. Similarly, proponents of creolisation have defended the possibility to hijack existing languages and categories through the radical encounter between different cultures resulting in the formation of something new out of something pre-existing (Bernabé et al., 1990; Gordon, 2014; Monahan, 2011; Sealey, 2020). Creolisation is thus an empowering appropriative movement consisting in "the strategic, unauthorised appropriation of symbols of power [...] which become empowering against their initial purpose" (Romberg, 2002). Indigenous scholarship also provides many examples of how Indigenous politics are actually appropriating key Western political concepts and their associated practices to imbue them with their own worldviews and make them work for their aspirations, thus gradually performing a conceptual decolonisation that can permeate the societies they live in as well as broader intellectual geographies (Coulthard, 2014; Lightfoot, 2021; Turner, 2006). Therefore, two main paths appear to open up in the task of decolonising political concepts. One pursues the reinstatement of erased languages and concepts different from those of the Western modern apparatus. Another focuses on re-signifying existing concepts in order to transform their conceptualisation and thus their impact on the world. Both are potentially complementary.

Naming is thus key in decoloniality. To name what has been excluded and negated. To adequately name what exists in order to reveal the coloniality of power and knowledge. Who gets to do the naming may be even more crucial. The Western modern rationality defining most of the concepts of predominant contemporary political thought constrains our thinking and forecloses the conditions of possibility. "Terminologies demarcate a field, politically and epistemologically. Names set up a field of power" (Trouillot, 1995: 115). When used uncritically, such terminologies may even condition and orient the cognition of anti-colonial or anti-racist discourses because of their underlying socio-political and structural implications. They may allow and reify a distortion of reality to comply with the matrix of coloniality, thus functioning as self-validating devices. This is why we need to lay bare how some conceptual tools are inadequately representing diverse experiences of reality and how they reproduce the conditions of coloniality. Adopting a decolonial attitude is to "critically review the entire arrangement of basic philosophical areas and concepts" (Maldonado-Torres, 2018: 124). It is also to articulate an alternative thinking starting from somewhere else than modernity/coloniality, from other ideas and imaginaries (whether or not we use the same tools). Therefore, in this volume and beyond, we aim to have an interdisciplinary and intercultural dialogue to challenge what Linda Tuhiwai Smith calls "research through imperial eyes" (1999: 42). We want to tackle the assumption that Western ideas and notions are the only rational ones, the only ones making sense of the world, of reality, of social life, and of human beings. This has to come hand in hand with a reversal of "hermeneutic injustice" (Alcoff, 2017) by which non-Western voices have been excluded from participating in the construction of the predominant concepts and terms that frame our collective world. The aim here is not only to include but to learn from these marginalised and silenced perspectives. Given the reciprocal incompleteness of cultures and knowledges (Santos, 2014), such an epistemological dialogue appears as the only possible path. Decolonising political concepts is precisely about birthing new constellations of meaning that acknowledge and embrace epistemological plurality and thus work against the illusion of objectivity and universality in Western conceptions.

At this point, the significance of addressing and re-thinking from a decolonial perspective the political concepts that make up contemporary political discourse should be palpable. They constitute windows through which we see the world, but whose frames can and do foreclose our possibilities and imaginaries. Some considerations on how 'political concepts' are generally understood throughout the volume seem necessary, especially in relation to the specificities of a decolonial endeavour. It is our collective understanding here that concepts are not merely political due to the fact that they are used to encapsulate, describe, or analyse matters concerning political life. What makes a concept 'political' is also the fact that it is deployed in political life itself in order to pursue and hopefully actualise particular agendas (Norberg, 2015). Conceptions thus can be – and often are – multiple in relation to the same concept. These differences in conceptualisation are oftentimes the result of either explicit or implicit diverging political ontologies, of attempts to influence and direct the 'reality-shaping' function that concepts perform, or of both. Contestation over the definition of political concepts is thus rarely the result of a purely intellectual exercise happening in a political vacuum. Their meanings are politicised and deeply embedded in the political realities in which they are produced and deployed.

A variety of actors, from academics to politicians or activists, mobilise the same concepts in diverging ways. Each of them advances a slightly to significantly different version of the same core idea in order to support their own claims and aspirations. This is how the same concepts have historically been articulated in progressive, conservative, democratic, authoritative, colonial, decolonial, oppressive, and emancipatory projects. Democracy stands as the banner of different political factions with radical variations in the values and institutions they ascribe to such an idea (Connolly, 1993; Skaaning, 2021). Sovereignty is claimed by both settler states and by some Indigenous communities in settler colonial contexts, although their conceptualisations are at times radically dissonant (Clavé-Mercier, 2022). Opposing sides have often faced – and even killed – each other due to their respective understandings of justice (Heywood, 2013). In the United States, the existence of two distinct views of freedom based on diverging moral and political beliefs has been argued to be one of the reasons for the existing profound social division (Lakoff, 2006). Political concepts thus appear contingent and historical. Active contestation around their definitions – as the ones illustrated in the volume's chapters in the case of 'freedom', 'autonomy', or 'indigeneity', for instance, is a testimony of their lack of universal, fixed, and intrinsic essences. In fact, most – if not all – political concepts are "essentially contested concepts" (Gallie, 1956) by the very fact that they deal with politics, a characteristic apparently heightened in the 20th and 21st centuries (Boromisza-Habashi, 2010). Analysing and conceptualising political life is thus, intrinsically, participating in it as well.

However, the deployment of political concepts in daily political discourse and practice and their analysis in academic fields are both marked by the same recurring trend towards depoliticisation:

> A problem that all political concepts share, and which the study of political concepts faces, is the urge to depoliticise. Political concepts are (relatively) settled meanings, and in studying political concepts there is an urge to set out a neutral (impartial, shared) language of political discourse, even or especially when dealing as politics does with contestability and deeply held differences. In addition there is a tendency to think that the study of political concepts has greater legitimacy and so trumps uses of political concepts by engaged practitioners.
>
> (Prokhovnik, 2007: 27)

In decolonising political concepts, we precisely seek to run counter to these depoliticisation efforts that more often than not enshrine Western conceptualisations as the universal and neutral essence of concepts that crucially define and shape our political realities. Such a task also requires us to question and re-configure our understanding of 'the political' or of '*political* concepts' as the semantic field they open up is neither clearly delimited nor are the genealogical lines simple or unambiguous. If we are to engage in a genuine epistemological dialogue with political ontologies that may significantly differ from those that have so far constituted the centre of contemporary political thought, this means that the borders of what is political may need to be pushed or blurred. Finally, decolonising political concepts is also impossible if we limit ourselves to the confines of the academic and intellectual spheres. As Raia Prokhovnik rightly stresses above, practitioners engaged in daily political practices are also key in conceptual definition and transformation. Given the considerable limitations placed on colonised peoples to access the well-guarded academic positions, as well as the significant weight that praxis has in the decolonial projects, the theoretical and conceptual agency of 'on-the-ground' practitioners needs to be adequately acknowledged and embraced if we are to decolonise political concepts.

As it stands, our collective endeavour thus joins in the efforts of conceptual history in that it similarly takes aim at the idea of clearly bounded, static, and stable concepts (Koselleck, 2002; Pernau & Sachsenmaier, 2016). We do agree with conceptual historians that concepts have histories and that their meanings mutate through time. However, we take these efforts further as we engage with voices and knowledges that are mainly side-lined in the resulting "histories of concepts". We do not mean here to adapt or implement Koselleck's theory of conceptual history, but rather to focus on and open a dialogue with conceptualisations that have been dismissed by a history largely written, understood, and conceptualised as European. To decolonise political concepts means to disrupt Western understandings, knowledge, and socio-political practices by unsettling embedded colonial relations. It means to give special attention to how coloniality has featured in the histories of different concepts and to make contributions to these histories made by equally valid theorising and practising subjects located beyond European modernity front and centre.

Finally, our use of labels such as 'West' or 'Western' throughout the volume warrants a final introductory consideration. First of all, 'Western political thought' is here understood as a "body of political, legal and social theory developed by European, American, Australian and New Zealand authors and practitioners from the beginning of the modern period in Europe to the present" (Ivison et al., 2000: 2). While it seems undeniable that Western political thought has been dominated by some form or variation of modern liberalism for some centuries now, it is however a varied and contested body. Neither is it diametrically opposed to non-Western forms of thought. The development of both Western and non-Western political thought has rarely – if ever – been hermetically sealed processes, especially in contexts where colonisation processes led (or forced) to mutual influence. The aim of this volume is definitely not to condemn theories and authors that can be considered 'Western', but to critically question them and to profitably introduce them into decolonisation debates. Some of our contributors' chapters actually advance a decolonisation of political concepts through discussions of theories and scholars that shaped the so-called 'Western' canon. In doing so, the contributions draw on different theoretical strands emerging from/within it, such as French Theory, the Frankfurt School or US-American Critical Race scholarship. In their discussion, it is shown that the particular theoretical frameworks developed by Western authors may also provide productive and emancipatory answers to the prevailing power structures. Thus, Western political thought should not be treated as a homogeneous corpus.

The same occurs with the 'West'. Throughout this introduction and this collection, our use of such a label does not stand for a particular geographical place or a specific site of knowledge production. Instead, it encapsulates a set of particular ontologies, epistemologies, worldviews, and subjectivities that are deeply intertwined with modernity/coloniality. The ideological West can thus be challenged from inside the geographical West or instead be predominant and defended outside of it. Borrowing from Yarimar Bonilla, the West is a project, not a place (2017: 334). A project in that it pursues the actualisation of a particular mode of being, of knowing, of deciding, of living collectively, etc. But also a project in the sense that it has never been totally and purely realised, it remains an aspiration. Much of this continued incompleteness rises from its constitutive relation with what would be the non-West, and from the continued existence of alternative modes of existence to those of the West. Indeed, as Hesse points out, "we should [...] consider the West as always already creolized by virtue of its modernity and coloniality" (2011: 41). However, 'West' and 'non-West' are still useful "short-hand for constellations of sources, issues, and methods of argumentation that [...] produce family resemblances and recognizable attributes" and thus "cannot be dispensed with, since they are forms of representation embedded in mythologies that anchor our understandings" (Gordon, 2014: 205). They are nonetheless terminologies to be adopted critically. The very least, as we did here, is to maintain an explicit acknowledgement of their intrinsic limitations, the main one being their obfuscation of the often intertwined constitution of what falls under these denominations.

Searching for a More Habitable Place: Decolonising Political Concepts

Throughout their contributions, the authors in this book challenge and reflect on colonial assumptions and implications that may be embedded in the categories that shape our surroundings and interactions. They aim to reveal and question how thought constructions influence the organisation of our socio-political realities, subjectivities, and lives. Zooming in on the crossroads between coloniality of knowledge and coloniality of power, they not only ask which perspectives became excluded through Western hegemony, but also how non-Western perspectives were at times shaped and appropriated by Western thought. Additionally, they also engage with alternative approaches deployed by colonised and marginalised political subjects throughout the world, pondering what decolonised political concepts might look like. It is also part of this project to recognise how different lines of oppression can intersect in the position of actors and groups in various settings and situations and, subsequently, lead to different understanding and experiences of political concepts. Taken together, the different contributions will reveal how the layers of coloniality play out in a way that always escapes the simplified coloniser/colonised binomy. The same conceptual constructs can simultaneously, alternatively, and contextually either articulate and facilitate oppressive and exploitative structures or encapsulate acts of resistance. This dimension of power inherent to concepts needs to be acknowledged, discussed, and challenged in order to understand and contest ongoing coloniality. We see these reflections on the naming of our complex realities and experiences, and its implications, as a necessary first step in devising genuinely postcolonial academic and political contexts. What is at stake here is the dismantlement of systems of oppression and marginalisation embedded in some of the central political concepts that still structure our thinking today. We argue that such a decolonising task is especially significant for those Western and Westernised scholars – such as most of us – wanting to meaningfully engage in decolonial work, as a way to deconstruct the tools and frameworks they have been given by the Westernised university.

The irony is not lost upon us that we are writing this volume from the Westernised university itself. The very same apparatus made of universities geographically and/or intellectually located in Western constellations of knowledge and structured around Western epistemological approaches. An academic world in which non-Western scholars still face significant structural disadvantages and suffer from various forms of exploitation, in which epistemic violence and suppression are still too often the common response to non-Western knowledges (Cupples & Grosfoguel, 2018). As epistemic institutions, universities have historically played a crucial role in depreciating the cultures and knowledges of colonised and Indigenous peoples. They too often continue to do so. However, as originally based on critical thinking and on questioning, the Westernised university simultaneously stands as a potential locus to confront and deconstruct the very structures of exclusion it is currently entangled with. We still believe in this potential and, via this volume, aim to add yet another spark in this direction in the wake of the decolonisation movements shaking universities across the world from Rhodes Must Fall to the many calls to decolonise curriculums across fields and disciplines. Therefore, while the Westernised university is still undoubtedly a key institution in the maintaining of coloniality in our contemporary times, it also contains the seeds of a challenge.

It is within this context that, in 2019, we organised a conference under the theme "Decolonising Political Concepts" at the University of Aberdeen. At the time, we both were doctoral candidates at the Centre for Citizenship, Civil Society and Rule of Law (CISRUL) and part of a PhD programme focused on the uses and contestations of political concepts in the 21st century funded by an H2020 Marie Skłodowska-Curie Action (grant agreement no.754326). On the back of a significant push in many social sciences and humanities disciplines for an active decentring of Western perspectives, we felt the need to bring decolonial and critical race insights into our research and thinking in order to confront the still too often unquestioned colonial genealogies, legacies, and loads of political concepts. Although decolonisation studies are undoubtedly becoming an established field, epistemology and, more concretely, political concepts are rarely addressed and unpacked as the prime focus. Even worse, they are at times assumed as given and unproblematic, thus potentially being counter-productive for otherwise critical and promisingly emancipatory projects. This book sets out to challenge this view.

Concepts beyond Borders

The book project, as is usual in such collective and long-haul endeavours, has seen some departures and some significant expansions. As a whole, its different chapters articulate efforts to cover diverse – but key – topics and geographies in the overall task of examining the relation between decolonisation and political concepts. Written by researchers located in various disciplines and geographical areas, this volume sketches a global picture of the multiple layers of coloniality through engagements with different local settings as well as different disciplinary lenses. The following pages open a room for exchange among philosophers, social scientists, and legal scholars. Such an interdisciplinary nature allows to understand the profound and manifold impact political concepts and their colonial heritage have on our actions and modes of thinking. Moreover, the different paths taken by the contributors in the common task at hand are not only a testimony to the volume's interdisciplinarity. They also highlight that concrete – and at times personal – circumstances, backgrounds, and contexts shape how coloniality is experienced, described, and challenged through political concepts. Voices from Africa, the Middle East, India, America, and Europe are directly or indirectly brought together to engage with a cast of contemporary political thought's most central political concepts such as freedom, the subject, agency, indigeneity, autonomy or sovereignty.

As a project in itself carried out by early-stage researchers, this volume consciously and proudly gives voice to young and emerging scholars with significant but not yet established contributions. Nonetheless, their reflections here are framed by a Preface and Afterword composed by established scholars who enjoy high reputations in their respective research areas and in the postcolonial/decolonial fields. This explicit commitment with early-stage researchers comes from the fact that students and young scholars are often spearheading the push for more rigorous and sustained efforts to decolonise intellectual traditions and academic fields. They are also key in generating new avenues of thought and in articulating innovative approaches and pathways in the face of established academic discourse. Moreover, as editors, we are convinced that decolonising political concepts, and more broadly decolonising academia, goes hand in hand with troubling the academic power dynamics and relations.

The present collection of contributions offers an entry point into the substantial task of decolonising contemporary political thought. Following the twofold nature of the "decolonial attitude", merging critique and creative generation, *Decolonising Political Concepts* is not circumscribed to exposing the ongoing coloniality enshrined in some contemporary political concepts. It also presents paths to re-think and/or overcome them into (a)venues for decolonial and postcolonial struggles. The discussion contained in its pages allies theoretical analyses with the exploration of concrete instances of political practice, two complementary lenses in the same critical examination of the relation between political constructs and coloniality/decoloniality. On the one hand, some of the contributions focus on specific political concepts such as agency, sovereignty, indigeneity, or freedom. Others examine concepts that might not traditionally be regarded as political concepts but prove to be of political and colonial weight like sexual violence or refuge. On the other hand, some contributors decided to dissect the context and processes of concept formation and deconstruction by reflecting on broader issues of knowledge production, history, and translation. Through these diverse contributions, the reader is guided along the different aspects and steps that a decolonisation of political concepts entails. After exposing the closures and limitations of the existing dominant epistemological framework, the colonial load present in some political categories and how we use them today is deconstructed before productively thinking through what decolonised political concepts might look like.

Moving Sideways, Looking Forward

Part I gathers contributions that address the issue at hand from a broad angle and in a marked theoretical and philosophical fashion. They mainly focus on the existing conditions of any intellectual endeavour in order to confront the coloniality embedded in the contemporary conceptual apparatus used to approach political life. Specifically, they seek to set the ground to challenge the relations of knowledge and power embedded in how the world surrounding us is commonly conceptualised. To do so, they expose how most of the Western (and liberal) epistemological coordinates reinforce coloniality in the predominant ways to approach and analyse reality. In doing so, they reveal the existing barriers to the decolonisation of political thinking contained in the currently predominant knowledge frameworks.

Under the title "Historicising History: A Critique Enabling View of History", Karim Barakat offers an account of the role historical analysis can play in decolonising existing political frameworks in political philosophy. By foregrounding the historicity of the preconditions for the formation of historical knowledge themselves, the predominant assumptions concerning what counts as historical knowledge are opened up to decolonial critique. Running counter to generally accepted universal views in political philosophy, history thus appears as a key lens in the local analysis of the appearance of specific political concepts and practices. Barakat thus proposes the historical method as a way to articulate critique without rooting it in assumed universal ideals often derived from modernity/coloniality, while still keeping relativism at bay through a clearer understanding of how historical social conditions contribute to the formation of historical accounts.

Laurencia Sáenz Benavides' "The Recalcitrance of White Ignorance" elaborates on Charles Mills' notion of "white ignorance" and its constitutive links to the coloniality of knowledge and the coloniality of power. Zeroing in on the "motivational group interest" aspect of white ignorance, the chapter points to the roots of an epistemological attitude that is not latent nor involuntary but decided and aggressive. Drawing from the works of several decolonial thinkers such as Audre Lorde, José Medina, or Sara Ahmed, Benavides posits that white ignorance is grounded on affective dimensions that intertwine knowledge and power and make its eradication extremely difficult. The chapter thus provides an understanding of white ignorance as a matrix of the coloniality residing in the Western conceptualisation of many political concepts. It also sustains that, in order to thrive, any decolonial project will need to contend with its conative and affective roots.

In his essay "The Idealised Subject of Freedom and the Refugee", Shahin Nasiri unpacks the highly normative overtones and universalist structure of the predominant conceptualisation of 'freedom'. Through an examination of the predominant discourses and theorisations of refugeehood, this essay charts the way the universalist discourse of freedom has been framed by the modern/colonial nexus of citizenship, statehood, and territorial sovereignty. As such, refugeehood is explored as a phenomenon and lived experience potentially disrupting the logic of the modern nation-state and its underlying colonial binomy of inclusion/exclusion. Nasiri, therefore, assesses refugeehood as a state of political (non)-subjectivity in which freedom is understood and practised in ways that call for a redefinition of freedom itself away from its modern/colonial roots.

Part II zooms in on the currently existing coloniality embedded in some political concepts. More concretely, it takes the ideas and experiences of agency and bodies as its primary focus. The two contributions contained in it thus discuss how coloniality is felt, performed, and enfleshed from an intersectional standpoint. To recognise ongoing colonial violence, to understand the struggles of liberation, also means to see the intersection of sexuality, race, class, and gender in concepts and their re-enactment.

In their chapter, "Politics without a Proper Locus: Political Agency between Action and Practice", Henrike Kohpeiß and Marie Wuth reflect upon the challenge of decolonising political agency. For this purpose, they engage with Hannah Arendt whose theory is one of the most influential references for understanding action and politics, but also part of a 'Western' tradition of political thinking. Arendt's specific concept of action comes with a theoretical framework, which cannot account for circumstances of violent domination under which some populations live and therefore runs the risk to exclude these parts of society from the field of political agency. With this observation, the idea of political agency itself is put under scrutiny and has to be reviewed with regard to alternative concepts. In contrast to Arendt's vocabulary, Saidiya Hartman's notion of practice aims to describe human activity under extreme circumstances of undermined autonomy and thereby suggests an altered view on politics, action and the construction of political subjectivity. Bringing Arendt into conversation with Hartman, as well as Fred Moten, the chapter portrays agency as a controversial political concept that marks a vantage point of political struggle.

Cecilia Cienfuegos Martínez's chapter articulates a critical entry to the thinking of sexual violence as political violence from a decolonial feminist perspective. The essay argues that sexual violence should be analysed not only as a repressive but also as a productive force, one imprinting on bodies a whole cosmology of power relations. Taking on an intersectional lens, its main proposition is that sexual violence is indeed an active construction of gendered and racialised distinctions that are not susceptible to universalisation under the paradigm of sexual difference. The first section of the chapter analyses the work of decolonial feminism to address how the imposition of the modern distinction between public and private as political/non-political becomes a fundamental construct for the depoliticisation of sexual violence. In the second section, the distinction between *body* and *flesh* is presented to epitomise the historical continuities of the violences of coloniality and sexual violence is dissected as a central element in the political distribution of power in modernity/coloniality.

Part III delves into the search for modes and languages of resistance that are not embedded in a colonial system of power. Across three chapters, our contributors articulate empirical case studies addressing different geographical contexts and various political movements in order to suggest possible routes for the decolonisation of political concepts. Their main concern lies in devising ways to articulate political claims and to subvert the language of resistance in accordance with an agenda of political liberation. They specifically engage with non-Western and Indigenous political thought and experiences, with colonised and subaltern peoples' deployments and conceptualisations of political concepts, to explore possible conceptual rearticulations and to discuss the possibilities of an emancipatory use of the coloniser's language and tools. Such dialogues, we argue, are necessary not only to devise what decolonised political concepts might look like, but also to find ways of living together that acknowledge and respect plurality. Meaningful decolonial academic and political contexts depend on it.

Written by Laura Galián, Chapter 6 reflects on the extent to which anti-authoritarian and anarchist collectives in the Southern Mediterranean have used language and translation to decolonise the political practice of anarchism. At the hands of these collectives, the translation of Western anarchist thought and concepts into Arabic is thus conceived as a political act working against the West and its languages as the centre of knowledge production. By examining the politics of language and linguistic practices of several anarchist and anti-authoritarian collectives in Lebanon and Morocco, Galián advances that language and translation politics are crucial to decolonisation as which languages are used necessarily delimit which knowledges are allowed and made central. As such, this essay invites a conversation on the plural histories and uses of concepts illuminated by the practices that concrete socio-political actors deploy to decolonise their conceptual apparatus.

"Decolonising Sovereignty and Reimagining Autonomy: Adivasi Assertions and Interpretations of Law", co-authored by Astha Saxena and Radhika

Chitkara, unpacks the idea of autonomy in the Indian context and examines its relationship with state sovereignty. Starting from the political and legal thought and practice of Indigenous communities such as the Adivasi and Indian forest-dwelling communities, the contributors explore possible rearticulations of the conceptualisation and practice of sovereignty based on the constitutional interpretations of the idea of autonomy by otherwise state-marginalised voices. Through an empirical discussion of the Pathalgarhi movement and the Niyamgiri protests, the essay argues for a decolonisation of sovereignty and autonomy towards an acknowledgement and embrace of political multiplicity. Otherwise, the authors argue, Indigenous movements in India may be trapped in modern/colonial political imaginations and configurations that not only make their aspirations for autonomy and self-determination impossible, but actually justify political and physical violence against them.

Rafael Verbuyst's chapter zeroes in on a recently reinvigorated decolonisation debate in post-apartheid South Africa. Within this particular context, Verbuyst posits that the concept of indigeneity constitutes an under-examined stumbling block to a successful decolonial endeavour. This essay thus proceeds to unpack the different perspectives on indigeneity being deployed by three main groups of actors: the marginalised Khoisan ethnic group, government officials, and scholars. Articulation theory provides the necessary tools to identify different understandings of indigeneity that are nonetheless often conflated by all stakeholders involved. Verbuyst argues that it is this very conflation that renders the recipients and aims of decolonisation unclear, a situation particularly frustrating for the Khoisan. In turn, the chapter points to the notion of "conflicting logics of autochthony" as an analytical lens better equipped to delineate competing positions within the politics of indigeneity. In this case, overcoming conceptual confusion and evidencing the politicisation of conceptual articulation are crucial steps in recentring marginalised voices and in generating productive pathways towards decolonisation.

Notes

1 More on these topics and movements can be found in Begum and Saini (2019), Cantzler (2021), Gómez-Barris (2017), Liboiron (2021), and Peters (2015).
2 For further discussion on these two strands of thought, their differences, and their relations, see the Special Issue of *Review* "From Postcolonial Studies to Decolonial Studies: Decolonizing Postcolonial Studies" (2006, vol.29 (2)) or Grosfoguel (2011), Mignolo (2011), and Bhambra (2014).

References

Alcoff, L. M. (2017). Philosophy and Philosophical Practice: Eurocentrism as an Epistemology of Ignorance. In I. Kidd, J. Medina, & G. Pohlhaus (Eds.), *The Routledge Handbook of Epistemic Injustice* (pp. 397–408). New York: Routledge.

Anghie, A. (2005). *Imperialism, Sovereignty, and the Making of International Law*. Cambridge: Cambridge University Press.

Anzaldúa, G. (2009). Speaking in Tongues: A Letter to Third World Women Writers. In A. Keating (Ed.), *The Gloria Anzaldúa Reader* (pp. 26–36). Durham: Duke University Press.

Barker, J. (Ed.) (2005). *Sovereignty Matters: Locations of Contestation and Possibility in Indigenous Struggles for Self-Determination*. Lincoln and London: University of Nebraska Press.

Bernabé, J., Chamoiseau, P., Confiant, R., & Taleb Khyar, M. (1990). In Praise of Creoleness. *Callaloo*, *13*(4), 886–909.

Begum, N., & Saini, R. (2019). Decolonising the Curriculum. *Political Studies Review*, *17*(2), 196–201.

Bendix, D., Müller, F., & Ziai, A. (Eds.) (2020). *Beyond the Master's Tools?: Decolonizing Knowledge Orders, Research Methods and Teaching*. London: Rowman & Littlefield.

Bhambra, G. (2014). Postcolonial and Decolonial Reconstructions. In G. Bhambra (Ed.), *Connected Sociologies* (pp. 117–140). London: Bloomsbury.

Bonilla, Y. (2015). *Non-Sovereign Futures: French Caribbean Politics in the Wake of Disenchantment*. Chicago & London: University of Chicago Press.

Bonilla, Y. (2017). Unsettling Sovereignty. *Cultural Anthropology*, *32*(3), 330–339.

Boromisza-Habashi, D. (2010). How are Political Concepts 'Essentially' Contested? *Language & Communication*, *30*(4), 276–284.

Cantzler, J. (2021). *Environmental Justice as Decolonization*. London and New York: Routledge.

Césaire, A. (2001). *Discourse on Colonialism*. New York: Monthly Review Press.

Chatterjee, P. (1993). *The Nation and Its Fragments: Colonial and Postcolonial Histories*. Princeton: Princeton University Press.

Clavé-Mercier, V. (2022). Politics of Sovereignty: Settler Resonance and Māori Resistance in Aotearoa/New Zealand. *Ethnopolitics* (Online First). https://www.tandfonline.com/doi/full/10.1080/17449057.2022.2096767

Connolly, W. E. (1993). *The Terms of Political Discourse*. Princeton: Princeton University Press.

Coulthard, G. (2014). *Red Skin, White Masks: Rejecting the Colonial Politics of Recognition*. Minneapolis: University of Minnesota Press.

Cupples, J., & Grosfoguel, R. (Eds.) (2018). *Unsettling Eurocentrism in the Westernized University*. London: Routledge.

Dussel, E. (1993). Eurocentrism and Modernity. *Boundary 2*, *20*(3), 65–76.

Dussel, E. (1995). *The Invention of the Americas: Eclipse of "the Other" and the Myth of Modernity*. New York: Continuum.

Escobar, A. (2020). *Pluriversal Politics: The Real and the Possible*. Durham: Duke University Press.

Fanon, F. (1967). *Toward the African Revolution: Political Essays*. New York: Grove Press.

Gallie, W. B. (1956). Essentially Contested Concepts. *Proceedings of the Aristotelian Society*, *56*, 167–198.

Gómez-Barris, M. (2017). *The Extractive Zone: Social Ecologies and Decolonial Perspectives*. Durham: Duke University Press.

Gordon, J. A. (2014). *Creolizing Political Theory: Reading Rousseau through Fanon*. New York: Fordham University Press.

Grosfoguel, R. (2007). The Epistemic Decolonial Turn: Beyond Political-Economy Paradigms. *Cultural Studies*, *21*(2–3) 211–223.

Grosfoguel, R. (2011). Decolonizing Post-Colonial Studies and Paradigms of Political-Economy: Transmodernity, Decolonial Thinking, and Global Coloniality. *Transmodernity*, 1(1). https://escholarship.org/uc/item/21k6t3fq

Grosfoguel, R. (2013). The Structure of Knowledge in Westernised Universities: Epistemic Racism/Sexism and the Four Genocides/Epistemicides. *Human Architecture: Journal of the Sociology of Self-knowledge*, *1*(1), 73–90.

Hartman, S. (1997). *Scenes of Subjection: Terror, Slavery, and Self-Making in Nineteenth Century America*. Oxford: Oxford University Press.

Hesse, B. (2011). Symptomatically Black: A Creolization of the Political. In F. Lionnet & S. Shih (Eds.), *The Creolization of Theory* (pp. 37–61). Durham: Duke University Press.

Heywood, A. (2013). *Politics*. London: Palgrave Macmillan.

Ivison, D., Patton, P., & Sanders, W. (2000). *Political Theory and the Rights of Indigenous Peoples*. Cambridge: Cambridge University Press.

Koselleck, R. (2002). *The Practice of Conceptual History: Timing History, Spacing Concepts*. Stanford: Stanford University Press.

Lakoff, G. (2006). *Whose Freedom?: The Battle Over America's Most Important Idea*. New York: Picador.

Liboiron, M. (2021). *Pollution is Colonialism*. Durham: Duke University Press.

Lightfoot, S. (2021). Decolonizing Self-Determination: Haudenosaunee Passports and Negotiated Sovereignty. *European Journal of International Relations*, *27*(4), 971–994.

Lorde, A. (2007). The Master's Tools Will Never Dismantle the Master's House. In A. Lorde (Ed.), *Sister Outsider: Essays and Speeches* (pp. 110–114). Berkeley: Crossing Press.

Maldonado-Torres, N. (2007). On the Coloniality of Being. *Cultural Studies*, *21*(2–3), 240–270.

Maldonado-Torres, N. (2011). Thinking through the Decolonial Turn: Post-Continental Interventions in Theory, Philosophy, and Critique—An Introduction, *Transmodernity*, 1(2). https://escholarship.org/uc/item/59w8j02x

Maldonado-Torres, N. (2016). Outline of Ten Theses on Coloniality and Decoloniality. Available at: https://fondation-frantzfanon.com/outline-of-ten-theses-on-coloniality-and-decoloniality/

Maldonado-Torres, N. (2018). The Decolonial Turn. In J. Poblete (Ed.), *New Approaches to Latin American Studies* (pp. 111–127). New York: Routledge.

Meghji, A. (2021). *Decolonizing Sociology: An Introduction*. Cambridge: Polity Press.

Menon, D. (2022). *Changing Theory: Concepts from the Global South*. London and New York: Routledge.

Mignolo, W. (2000). *Local Histories/Global Designs: Coloniality, Subaltern Knowledges, and Border Thinking*. Princeton: Princeton University Press.

Mignolo, W. (2011). Epistemic Disobedience and the Decolonial Option: A Manifesto. *Transmodernity*, 1(2). https://escholarship.org/uc/item/62j3w283

Mignolo, W. (2021). *The Politics of Decolonial Investigations*. Durham: Duke University Press.

Monahan, M. (2011). *The Creolizing Subject: Race, Reason, and the Politics of Purity*. New York: Fordham University Press.

Norberg, J. (2015). Concepts, Political. In M. T. Gibbons (Ed.), *The Encyclopedia of Political Thought*. Oxford: Blackwell.

Pagden, A. (1995). *Lords of All the World: Ideologies of Empire in Spain, Britain and France c.1500-c.1800*. New Haven: Yale University Press.

Peters, M. (2015). Why is My Curriculum White? *Educational Philosophy and Theory*, *47*(7), 641–646.

Pernau, M., & Sachsenmaier, D. (Eds.) (2016). *Global Conceptual History. A Reader*. London and Oxford: Bloomsbury.

Prokhovnik, R. (2007). *Sovereignties: Contemporary Theory and Practice*. Basingstoke: Palgrave Macmillan.

Quijano, A. (1992). Colonialidad del poder, Eurocentrismo y América Latina. *Perú Indígena*, *13*(29), 11–20.

Quijano, A. (2000). Colonialidad del poder y clasificación social. *Journal of World-Systems Research*, *2*, 342–386.

Quijano, A. (2007). Coloniality and Modernity/Rationality. *Cultural Studies*, *21*(2–3), 168–178.

Romberg, R. (2002). Revisiting Creolization. Available at: https://www.sas.upenn.edu/folklore/center/ConferenceArchive/voiceover/creolization.html

Said, E. (1993). *Culture and Imperialism*. London: Chatto and Windus.

Santos, B. (2014). *Epistemologies of the South: Justice Against Epistemicide*. New York: Routledge.

Sealey, K. (2020). *Creolizing the Nation*. Evanston: Northwestern University Press.

Shaw, K. (2008). *Indigeneity and Political Theory: Sovereignty and the Limits of the Political*. London: Routledge.

Shilliam, R. (2021). *Decolonizing Politics: An Introduction*. Cambridge: Polity Press.

Simpson, A. (2014). *Mohawk Interruptus: Political Life across the Borders of Settler States*. Durham: Duke University Press.

Skaaning, S. (2021). Democracy: Contested Concept with a Common Core. In G. Crawford & A. Abdulai (Eds.), *Research Handbook on Democracy and Development* (pp. 27–44). Chelthenham: Edward Elgar Publishing.

Smith, L. T. (1999). *Decolonizing Methodologies: Research and Indigenous Peoples*. London: Zed Books.

Trouillot, M. (1995). *Silencing the Past. Power and the Production of History*. Boston: Beacon Press.

Turner, D. (2006). *This is Not a Peace Pipe: Towards a Critical Indigenous Philosophy*. Toronto: University of Toronto Press.

Walker, R. (1987). *Nga tau Tohetohe. Years of Anger*. Auckland: Penguin.

Wiredu, K. (1995). *Conceptual Decolonization in African Philosophy: Four Essays*. Ibadan: Hope Publications.

Wiredu, K. (1996). *Cultural Universals and Particulars: An African Perspective*. Bloomington and Indianapolis: Indiana University Press.

Wood, D. A. (2020). *Epistemic Decolonization: A Critical Investigation into the Anticolonial Politics of Knowledge*. London: Palgrave Macmillan.

Wynter, S. (1995). 1492: A new World View. In V. Hyatt & R. Nettleford (Eds.), *Race, Discourse, and the Origin of the Americas: A New World View* (pp. 5–57). Washington: Smithsonian Institution Press.

Wynter, S. (2003). Unsettling the Coloniality of Being/Power/Truth/Freedom: Towards the Human, after Man, Its Overrepresentation – An Argument. *The New Centennial Review*, *3*(3), 257–337.

Part I

Decolonial Horizons

Revealing the Coloniality of Knowledge and Power

1 Historicising History

A Critique Enabling View of History

Karim Barakat

The extent to which decolonial theory has been committed to a form of epistemological and ethical relativism has been a subject of thorough scrutiny.[1] Yet, to maintain the power of the decolonial critique, it seems necessary that we reject a full-fledged relativism that would undermine the possibility of pushing for a change not emanating from Eurocentric systems of knowledge. Instead, a historical analysis of the effects of colonialism is required, in addition to making use of cultural resources of historically marginalised communities. Decolonising politics must, consequently, examine how we conceive the context that has constituted our political concepts. If we recognise that decolonial critique is grounded in historical analysis as we must, this introduces the need for a critical scrutiny of the conditions that make historical analysis possible. In other words, if political problems involve specificity as an outcome of historical analysis, does not the framing of political problems also raise the issue of how we ought to view historical knowledge in the first place? As I show in this chapter, this is primarily due to the fact that historical knowledge is constructed, and how this construction takes place determines to a large extent how we engage critically already existing political concepts and structures. The question that follows is how history ought to inform us when it comes to political critique.[2] This raises at least two problems. First, we need to establish a reliable and objective standard for what counts as historical knowledge. Second, this standard should not reduce critique beginning from historical knowledge to a form of universalism, and it should make room for the possibility of cross-cultural communication, without pushing for universal prescriptive commitments.[3]

This chapter will, therefore, focus on two problems that relate to exploring the role one can assign to history in politics. The first follows from the prescriptive aspect of political philosophy, which has predominantly entailed introducing universal conclusions that go beyond local phenomena. Under this conception of political critique, however, it becomes unclear how historiography, as a discipline concerned with specific analyses of individual events can offer any insight. Conversely, the excessive reliance on history in

DOI: 10.4324/9781003293460-3

political prescription could collapse into a form of relativism that rejects any kind of overlap between different groups and communities. The problematic outcome that could follow is a steadfast conservatism that rejects any social change under the guise of authenticity and appealing to tradition. Recognising that historical knowledge can offer insights when it comes to political and social criticism ought to involve introducing a conception of historical knowledge that does not collapse into relativism. Instead, historical analysis must be able to take into account the possibility of the emergence of a plurality of social phenomena that are not violently reduced to universal forms of analyses, while simultaneously making room for critique that transcends the locality of social practices. Critique can only follow from going beyond the confines of a limited context for social practices and from engaging the other. Social criticism, therefore, requires tapping into resources that could offer grounds for generality, though not universality.

The second problem involves the question of which historical accounts should be informing our political conclusions. As I show in the upcoming sections, the problem of the relativity or objectivity of historiography has been central to philosophical inquiries focusing on historical knowledge. If we accept that a local and contextualised, and thus historical, understanding of political problems can offer insights when it comes to determining the kind of political transformations needed, we are still encumbered by the task of specifying which historical accounts constitute a legitimate basis with which we can begin. In other words, what we need is some method or a set of criteria that enable us to adjudicate between different historical views, especially at a time when fictitious historical narratives are rampant.

In what follows, I will not attempt to offer a resolution to the problem of the objectivity of knowledge, as that arduous task goes beyond the scope of the chapter. Instead, I invoke some notions from the philosophy of history and historiography to show how historical accounts can avoid relativism while being a useful tool for political critique. I begin by shedding light on the problem historically by presenting some of the unacceptable consequences that follow from endorsing a realist view with respect to historical knowledge. The first section of the chapter briefly discusses the contributions of the empirical, transcendental, and postmodern lines of argument. I then show how we can overcome the challenge of a relativistic view of historical knowledge, while arguing that political thought ought to focus on social conditions that contribute to the emergence of historical events. In order to establish my conclusion, I seek to draw bridges between political philosophy and the philosophy of history through examining the kind of explanation history offers. I argue that we can understand historical explanation for political problems in terms of social conditions that offer partial sufficient conditions of possibility. This implies that unique historical conditions could result in the emergence of similar problems in different social contexts, which allows for invoking general prescriptive statements.

Models of Historiography

The Empirical Model

The relationship between philosophy and history has been a subject of a perennial debate that stretches over different periods of the history of philosophy. Aristotle had famously claimed that history is less philosophical than poetry (Aristotle, 2009: 1463–4), a view that dominated until the appearance of the notion of universal history in seventeenth-century European political thought (Zammito, 2009: 65). The insistence on imposing a universal character on history in the seventeenth, eighteenth, and nineteenth centuries, however, did not offer an adequate conception of how actual specific historical explanation can be informative. This antagonistic attitude towards historical analysis persisted in some form during the twentieth century. In his 1949 article "Political Philosophy and History", Leo Strauss argues that "political philosophy is not a historical discipline" (Strauss, 1949: 30). For Strauss, independently of any historical accounts, political philosophy raises "questions of the nature of political things and of the best, or the just, political order" (Strauss, 1949: 30). Introducing history to political philosophy would be tantamount to confusing specific features of political thought that are the product of a specific time with the more universal question of the nature of political things.

Whereas the Straussian line of argument problematically dismisses any possible role for historical analysis in political philosophy, his characterisation allows for shedding light on the problem at hand. By placing too much emphasis on the local nature of historical analysis, we run the risk of rejecting critique altogether and slipping into a problematic relativism. For Strauss, it seems that in the absence of some universal framework, critique can no longer be grounded in a rational capacity that is able to transcend differences. Accordingly, political critique presupposes that one recognises some generality to the political condition, which can be used as a basis for changing present practices.

The historiographic attitude towards philosophy has also been uncharitable. Ever since its inception with Leopold Ranke in the nineteenth century, scientific historiography has been anti-philosophical in principle. As Thomas Gil shows, Ranke's method involved emphasising the priority of facts, while avoiding generalisations and speculations in which philosophers engage. Moreover, Ranke's historiography stresses the uniqueness and individuality of historical events, while rejecting the overall notion of progress and teleology as inherent in history, a view dominant in German philosophy during the nineteenth century (Gil, 2009: 384–5). Yet, though Ranke departed from core commitments philosophers of his time held, he still insisted on developing a *scientific* historiography, despite it often falling short of offering objective criteria. As Zammito argues, Ranke's view of scientific historiography was influenced by numerous political commitments, in addition to various other implicit values that brought into the foreground the problem of how historiography can be value-laden and scientific at the same time (Zammito, 2009: 66). Furthermore, in the absence of a clear standard for objectivity, it is unclear what role historical analysis can play in political critique if it is to be motivated by already existing political values. These latent commitments left Ranke's view susceptible to the attacks of empiricists, who sought to identify a unique method for ordering historiographic knowledge.

The empirical approach the logical positivists pursued set as its goal arriving at a unified conception of scientific explanation (Brzechczyn, 2009: 416). If history is to be recognised as a form of knowledge, then it ought to follow the same structure we find in the explanation of natural phenomena. Logical positivists attempted to put forward a framework involving necessary and sufficient conditions to account for all causation in scientific fields.[4] Historical conditions can together constitute sufficient grounds for the appearance of historical events. Some historical conditions, moreover, may simply be necessary without being enough to bring about those events. Yet, as Tucker argues, reducing causation to necessary and sufficient conditions fell short of accounting for explanation, for historical causes are often neither necessary nor sufficient. Moreover, necessary and sufficient conditions do not have to be causal in nature, and therefore, the goal of identifying the explanation of a historical event would not be met (Tucker, 2009: 100–2).

Soon after, Carl Hempel presented his own extension of the positivist view to historiography, inaugurating what came to be known as Logical Empiricism. In his essay "The Function of General Laws in History", Hempel extended his argument for the model of a covering law to conceive of explanation in the natural sciences to history. According to this framework, observed individual events instantiate general laws. But it is the appeal to a general law that would ground scientific explanation. Hempel argues that one can arrive at historical laws by identifying patterns in history in a similar way we induce laws in the natural sciences, where actual events are nothing but instances of them (Hempel, 1942; D'oro, 2009: 143).

This view, however, was subject to a number of criticisms, one of which rightly criticised collapsing the explanations we find in the natural sciences with those of historiography. Peter Kosso argues that identifying general laws in history seems to be an arduous task since historical events depend on too many factors, many of which are influenced by human beings and thus on the spontaneous actions of agents. Unlike natural events, historical events are in fact unique, which makes it impossible to organise them under any number of general laws involving necessity (Kosso, 2009: 23–4). The empirical line of argument pushing for organising historical knowledge on the model of the natural sciences reached its pinnacle with the view developed by Hempel. Nonetheless, these views centred on marking out law-like regularity in a discipline where law-like patterns appear to be scarce. More importantly, the formal model Hempel proposes is not amenable to the structure of a narrative, and is therefore at the very least too limited to yield an account of historical explanation (Kosso, 2009: 23–4).

As we will see in the next section, introducing problems concerning the use of language into our view of historiography renders the problem even more thorny. The next section will shed light on concerns that follow from the linguistic turn in philosophy that occurred in the early twentieth century, before examining an alternative approach committed to the objectivity of historical knowledge following the Kantian tradition. These transcendental views did not invoke a model unifying natural and historical knowledge, but instead emphasised a necessary role for *a priori* categories in organising such explanations.

The Constructivist Model

The previous section articulated problems that followed from attempting to stave off relativism in historiography by reducing historical knowledge to law-like regularities. The proposed empirical model was unable to account for a large variety of historical accounts. Moreover, from a political standpoint, the empirical approach did not acknowledge the uniqueness of historical events. Alternatively, this section aims at showing the justifications behind moving towards transcendental accounts, while examining the political consequences of accepting these views in terms of the possibility of historical critique. Whereas the empirical approach to the philosophy of history emphasised experience in deriving general laws, views informed by Kant's transcendental idealism were also introducing a basis for the objectivity of historiographic knowledge, one that presided in the subject of experience.

To motivate the transition to this approach, it is useful to briefly look at the role Arthur Danto's analysis of language played in showing why it is problematic to be committed to a straightforward realism with respect to historical explanation as the empiricists maintained. As Ankersmit argues, Danto had shown that certain statements about the past can only be true conditionally (Danto, 2007: 152–3). For instance, the statement "I am in Beirut today" can only be true today. Moreover, historians themselves devise new concepts and terms to refer to past phenomena, rendering the truth of statements employing such terms dependent on context and history. To clarify, statements distinguishing past forms of colonialism can only be true after reflecting on a past era in a way that was not visible for those living during that time. For Danto, it is, therefore, the historian's task to introduce a context to clarify statements that were taken to be true in the past. The historian, therefore, provides the framework, a *representation* of a past world, that makes it possible for statements or narratives to be true or false about the past (Ankersmit, 2009: 202–3; Danto, 2007).

With these problems, it becomes difficult to see how an empirical approach to history could resolve issues of objectivity as it fails to take into account the temporal conundrums Danto articulates. Danto, however, derives from the conditional truth of historical statements a metaphysical conclusion or a "descriptive metaphysics" of historiography (Ankersmit, 2009: 202). In other words, the conditional truth of historiographic accounts reveals for us the reality to which narratives refer. Yet, Danto's account is still permeable to objections that motivate the transition to transcendental approaches. Ankersmit argues that Danto's view is improved with the modifications introduced by Baumgartner, who criticises Danto's position for its commitment to the reality of implausible notions (Ankersmit, 2009: 204). Under Danto's representation, statements about the Middle Ages, for instance, are true due to their characterisation of a *real* epoch in the world. Baumgartner, in contrast, chooses to shift the locus of truth from the world to the subject. Following a transcendental approach, Baumgartner argues that the representation provided by the historian is not indicative of a real world that matches our historical concepts. Instead, representation shows the role the subject plays in *constructing* historical accounts.

Baumgartner, therefore, endorses a Kantian transcendental position and argues that the construction of narratives does not offer insight into the actual state of affairs in the world, but rather informs us of the conditions that render a historical narrative intelligible for us. As Ankersmit words it, "story-telling and action require the capacity to reduce the manifold of the relevant contextual data to one single perspective from which either the narrative is organized or action is decided upon" (Ankersmit, 2009: 202). Narratives are, therefore, primarily constructed rather than experienced.[5] For Baumgartner, narratives are intelligible on the condition that they involve the necessary categories of unity and continuity, both of which are aspects provided by us.

A priori rules, therefore, determine what constitutes a *coherent* and *intelligible* historical narrative and distinguishes it from an incoherent string of events. Moreover, given that these rules that make narratives possible are independent of experience, they are universal in nature, even though the narratives themselves treat particular and individual entities. Accordingly, historical narratives can offer explanations for local phenomena while simultaneously following objective universal guidelines of continuity and unity. This understanding of history is also shared by Rickert who identifies a distinct role for each of history and philosophy. As a transcendental idealist, and in opposition to the empirical tendency discussed in the previous section, Rickert distinguishes between knowledge of the natural world and knowledge of history. Rickert, thus, argues that the objectivity of nature and history are made possible by distinct *a priori* categories, where the task of philosophy is to investigate these categories that make both forms of knowledge possible. For Rickert, historiography, on the one hand, is an empirical science of causal processes which is involved with analysing unique and individual events. Philosophy, on the other hand, investigates the *a priori* conditions of possibility for deriving conceptual generalisations (D'Amico, 2009: 244–6; Rickert, 1962). While rejecting a unified model of explanation surpasses the problems that follow from endorsing Hempel's covering law view, accepting these distinct roles for history and philosophy takes us back to our initial problem. If the philosophy of history is merely concerned with the *a priori* conditions that make historical knowledge possible, then it remains unclear what insight historical knowledge, which is empirical in nature, can offer in relation to philosophical critique, especially one that relies on historical accounts.

The conclusions Baumgartner draws also result in problematic implications with respect to whether history can provide any critical prescriptive assertions that are general in scope. His reliance on very basic categories renders it difficult to move beyond the particular nature of historical analysis to derive general norms. This leads to an implausible implication if we recognise that political problems are often constituted by historical circumstances. Historical explanations, especially ones that rely on narratives, invoke a local and particular analysis emphasising the role of contingency in the emergence of social phenomena. Such local analysis cannot be used to derive generalised claims about political problems. Under this view, it is no longer clear how historical knowledge can contribute to engaging in social critique.

The turn to transcendental philosophy affords an advantage over empirical views by recognising that historical analysis cannot be reduced to true or false judgements about statements concerning the past, but must instead acknowledge that the truth of such statements requires introducing a wider context. The strong representational element involved requires a more nuanced view of what counts as good and bad historiographic narratives, one that is not simply committed to the metaphysical reality of temporal states of affairs. Yet, by grounding historical knowledge in *a priori* notions, these views are unable to accommodate theoretically how one can use history for the purposes of critique. I argue later that this is primarily due to very neatly separating the philosophical task, as one concerned with an *a priori* analysis of conditions of possibility of knowledge, from empirically engaging the social and historical context. Before expounding on the alternative, I put forward, I will first offer a brief account of the turn to literary theory, along with the strong re-emergence of the problem of relativism in the philosophy of history.

The Postmodern Historiographic Model and the Charge of Relativism

The charge of relativism against historiography had already been widely articulated throughout the twentieth century. Perhaps Maurice Mandelbaum states the relativist arguments most clearly in *The Problem of Historical Knowledge: An Answer to Relativism* before arguing against the relativity of historical knowledge. Mandelbaum contends that the relativist position makes two distinct, but related, arguments. First, given the historian's inability to capture historical events the way they occurred, it is impossible for historiography to be accurate in any meaningful sense. Second, it follows that all historical views are value-laden and do not offer knowledge of the world as it is (Mandelbaum, 1938: 36).

The transcendental views outlined in the former section can respond to the first argument by showing that the historian works on a representation that is constructed by her. Consequently, one need not assume that we are uncovering past events as they really occurred by introducing a historical representation. Yet, the transcendental approach to historiography was still unable to specify in what sense knowledge can be both objective and value-laden. The argument from relativism, moreover, resurfaced strongly with the failure of Hempel's covering law model. The realist historiographic standpoint was subjected to a critique that extended the already existing relativist argument of historiography's inability to arrive at knowledge of the past as it is. The works of Foucault emphasising the historicity of epistemes, which determine what knowledge is possible along with Kuhn's view of normal science and revolution, cast further doubt on the view Hempel sought to entrench (Zammito, 2009: 69; Foucault, 2002, 2012; Kuhn, 2012). Yet, it is only with postmodern interventions emerging from literary theory that the relativist viewpoint in historiography acquired further force. The failure to assimilate historical knowledge with that of scientific knowledge, and the inability to put forward a single model that can encapsulate all historiographic explanations opened the space for a critique focusing on the plurality of methods used by the historian to generate a narrative. Perhaps the commitment to relativism in historiography culminated with the work of Hayden White, who focused on the plethora of tools used to make narratives in historical accounts compelling (White, 1975: 29). White argues that narratives are more invented than found. Historical narratives according to White have the aim of convincing the reader, just as a novel would (Ankersmit, 2009: 205). As Zammito points out, past events according to White have no order themselves and are only ordered through linguistic tools. There is no knowledge of the world without language, and thus there cannot be knowledge of the past that will not be encumbered by the relativist challenge resulting from the use of language (White, 2010; Zammito, 2009: 77).

For White, therefore, all historical accounts will be dependent on a paradigm of already accepted beliefs that guide the historian in terms of how to structure and shape the manifold of historical facts. White argues that metahistorical paradigms are always active in a historian's work, rendering the work largely the product of subjective elements (White, 1975). Ankersmit shows how White was initially committed to what appears like transcendental categories that make historiographic knowledge possible. Yet, unlike Baumgartner, White does not argue that these are *all* necessary for a historiographic narrative to make sense. Instead, these are optional tools the historian can use to put forward her point forcefully (Ankersmit, 2009: 206–7).

The consequence of this position, according to White, is that there can be "equally plausible, alternative, and even contradictory stories" (White, 2010: 231). Moreover, it is not possible to select better or worse alternatives between these accounts. It follows that there is no single model that can be relied upon in order to offer an explanation in history. Historical narratives often make use of a multitude of rhetorical tools to convey certain accounts, and these could produce contradictory narratives. Furthermore, as Ankersmit argues, White takes this analysis to its limit. By eliminating a unifying model, or an organising set of models, White argues that the historian is engaged in a radically free project of organising the evidence at hand. Underlying political and moral commitments always contribute to how the historian orders events of the past, an order that is invented (Ankersmit, 2009: 207). By arguing against the need for clear explanatory models that house all kinds of historical explanations, White gives further impetus to the problem of relativism in the philosophy of historiography.

The relativism emerging from White's view of history should be clear at this point. As Newall points out, the sceptical relativistic challenge to historiographic objectivity involves the assertion that alternative historiographic views cannot be adjudicated based on a conception of truth, a view to which White is certainly committed (Newall, 2009: 173). The political stakes for this consequence are also troubling. As I have argued earlier, critique requires that we are able to transcend social and historical factors in order to push for desirable change. Yet, identifying what change is desirable and what is not must follow from being able to distinguish true from fictitious historical accounts, which allows us to reject certain political outcomes and accept others. In the absence of a standard to determine truth from falsity, the use of history in politics ceases to be insightful, and its use becomes merely polemical. This is the conclusion that follows inevitably from White's account, as he insists that historical views emerge out of pre-existing political and moral commitments. Ankersmit presents a similar argument against White's view, where metahistory could explain why we find certain historical views compelling, yet it does not explain what views we *should* find compelling or which we are rationally justified in accepting (Ankersmit, 2009: 207). Under White's account, it follows that historical narratives cannot be used to revise pre-existing values, but rather come to reinforce entrenched views.

Social Conditions and Contingency

The previous section argued that, taken to its logical conclusion, the relativist view undermines the justification behind any political critique. The relativist standpoint renders historical knowledge impossible and accordingly makes critique based on historical analysis entirely reducible to one's already existing values, thus bolstering a conservative attitude. From a decolonial standpoint, critiquing pre-existing colonial frameworks is thus stripped from any legitimate basis upon which it can stand. Moreover, one can no longer argue forcefully for any decolonial view and critique of pre-existing values and social practices becomes impossible.

Avoiding those consequences requires achieving two goals; first, it must be shown that historical knowledge can be possible without appealing to a problematic realism. Second, the philosophical task of specifying the conditions that make historical knowledge possible cannot be insulated from historical analysis, and philosophical critique informed by historical accounts must avoid appealing to reductive universal prescriptions.

The models I have focused on so far provide three alternatives for dealing with historical knowledge. The empirical approach attempted to avoid altogether the problem of the inherent normativity and value-ladenness of historical accounts, thus reducing historical knowledge to true statements about the past. Despite presenting us with clear criteria for determining what counts as historical knowledge, this model fails to account for the act of constructing historical accounts and does not take into consideration the historicity of our criteria for what counts as legitimate historical knowledge.

Alternatively, the transcendental views resolve several of these problems by rejecting the realism we find among empiricists. For transcendentalists, the historical world is constructed by subjects that structure the manifold of historical data according to clear rules that determine sound from unsound historical accounts and determine the conditions of intelligibility of narrative. In a manner similar to the empirical line of argument, however, transcendental views are committed to the universality and necessity of our criteria regarding historical knowledge, thus rejecting their historicity. The political consequence is that philosophical inquiry ends up being divorced from the real world, and it becomes unclear how historical knowledge, which is contingent in nature, can permeate the necessary and universal realm of philosophy offering insight and enabling critique. Moreover, I argue in the next section that specifying necessary and universal, and thus ahistorical, criteria for the possibility of historical knowledge, can be invoked to exclude already marginalised narratives. It follows that decolonial critique must not operate only on already existing historical views, but must also engage the conceptual preconditions that are considered to be necessary for those accounts.

Finally, postmodern views resolve several of the above-mentioned problems. For the postmodern historian, there is no reality to which historical accounts are referring, for the historian is heavily involved in constructing the narrative using whatever tools are more compelling to the reader. Furthermore, the criteria that determine what historical accounts should be accepted and which do not satisfy the conditions for knowledge are themselves shifting, and do not offer ahistorical strict guidelines. But this last advantage also leads to the main disadvantage of such views, which have done away altogether with any criteria for the validity of historical accounts. We are, thus, left with a pernicious relativism that neither offers a ground for social criticism, nor allows for distinguishing well-justified historical research from compelling fiction, and could collapse into an unwavering conservatism.

Social Conditions as Conditions of Possibility of Historical Knowledge

Historicising the conditions that make historical knowledge possible must begin by recognising them as the outcome of social factors as opposed to fixed conditions of possibility rooted in the subject. Rather than invoking transcendental elements, these conditions become the outcome of a multiplicity of social and political factors and must themselves be subject to continuous rational critique for the sake of expanding inclusion. Social conditions, moreover, not only make historical knowledge possible, but they are also the preconditions for the emergence of political problems. Contextualising the appearance of political problems renders the application of universal prescriptions at the very least problematic, independently of identifying the historical process that brought about those problems. Yet, the absence of a universal prescription need not involve breaking down societies into isolated cultural spheres and thus endorsing a relativist standpoint. I argue, therefore, that the historicity of the conditions of possibility of historical knowledge still allows for general prescriptions, though not universal ones.

The inevitable conclusion of historicising the conditions of possibility of historical knowledge renders them contingent rather than necessary. It follows that the historian is certainly involved in offering a historical representation as Danto emphasises, yet this representation need not be grounded in ahistorical concepts, and it is not uncovering a past exactly as it occurred. The historical approach I argue for, thus, replaces the transcendental universal and necessary categories with ones that are shifting and historically specific. Recognising this changing nature of the preconditions of historical knowledge, thus, also allows for uncovering the structures of inclusion and exclusion at work in offering historical explanations. Historically, these systems have often served the purpose of reinforcing narratives of superiority, and they have justified colonial practices in many cases. The often-cited example Edward Said provides of Marx's use of history to justify British colonialism in India is a clear illustration of this point. Having determined that historical explanation must follow a universal teleological path that identifies social and economic progress as moving from feudalism through capitalism before arriving at communist liberation, Marx saw in British colonialism a force that would modernise Indian society and thus bring it closer to the conception of freedom he had envisaged (Said, 2004: 153). Providing a critique of this analysis cannot only pursue disputing the endorsed historical narrative. It must also challenge presuppositions concerning what counts as a historical explanation, such as teleological determinism as a necessary criterion for sound historical explanations. Critiquing colonial politics, therefore, must begin not only by acknowledging other historical accounts but also by critically engaging already existing preconditions that determine what counts as intelligible or unintelligible historical views. At the same time, such a critical approach must maintain a minimal conception of objectivity, thus avoiding the collapse into a relativism that renders all alternative forms equally acceptable.

This leads to several additional advantages. If philosophy is concerned with examining contingent social conditions instead of universal and necessary ones, it is no longer working in a realm that is insulated from empirical knowledge, and it can make use of historiography to transform its own subject matter. In other words, the social conditions that philosophy is concerned with analysing and critiquing can now be scrutinised by referring to established historical narratives. Introducing a historical account of the social conditions themselves would be useful in two ways. First, understanding the historical developments that led to the appearance of such social conditions allows us to have a fuller grasp of present political problems. To illustrate, developing an adequate understanding of the economic collapse that a country like Lebanon is undergoing today, for instance, can only follow from analysing the historical conditions that produced today's problems. These involve recognising the effects of a long-term civil war and a corrupt elite that has employed the country's resources for accumulating wealth and maintaining political capital. But, importantly as well, one must highlight the social consequences of a political setup that institutionalised religious and sectarian divisions in the state during colonial times. Alternative conceptual tools that make possible explanations in terms of how it is that institutional structures have enabled keeping a population in check despite the collapse of the infrastructure and economy are needed for conceptualising more fully the political problems at hand. The use of categories such as modern institutional structures in historical explanation is one that only became possible throughout the second half of the twentieth century with the work of Michel Foucault among others. Understanding the political problems specific to Lebanon, therefore, requires a historical analysis that recognises the roles of specific agents in leading to the present crisis, but that is also sensitive to the use of categories that go beyond mere agential responsibility, categories that broaden what counts as social and historical explanations. It follows that not only are the social conditions that produce political problems contingent but the criteria that determine what counts as historical explanation is itself subject to historical change and contingency. Despite these criteria being themselves preconditions for offering historical explanations, one has to acknowledge their own historicity, and their transformation in accordance with changing conditions over time. Consequently, it would have been impossible to put forward an intelligible account of the effects of colonial practices in Lebanon at a time during which epistemic criteria were blind to the role of state institutions in reinforcing social divisions.

Second, a historical understanding of how the problems arose enables us to understand the contingency of specific conditions that brought our situation about.[6] A historical account, therefore, offers the means to determine what conditions we can change, and how things could have been different. A decolonial critique of politics, therefore, can only arise from a historical analysis that unveils the heterogenous social processes that lead to the appearance of specific social problems. This historical analysis itself, however, needs to be aware of the complex ways in which power relations permeate any field of knowledge. Consequently, a sober critical engagement with the conditions that make historical knowledge possible is necessary for decolonising how we conceive of political problems as well.

Objectivity in Historical Explanation and General Prescription

Whereas the previous section introduced a conception of the conditions of possibility of historical knowledge as social and historical in nature, this section aims at providing a minimal account of objectivity that would render critique possible. Avoiding the relativism that postmodern discourse seems to endorse requires that we offer some standards that allow for objectivity in historical knowledge.[7] One way to achieve that goal is to understand the social context in terms of sufficient conditions that allow for the emergence of historical events. To clarify, these conditions need not be causal, and thus they can only provide *partial* sufficiency. Given that historical knowledge partly requires offering a causal explanation, these conditions will provide the background that render historical causes intelligible, yet they will not themselves constitute a full explanation for historical phenomena. In addition, social conditions are themselves historical rather than necessary and universal. Moreover, identifying these conditions involves a clear element of construction by the historian. They are not mirroring the past as it was, nor are they metaphysical in nature. The view of objectivity under this reading will have to deviate from thinking of historical knowledge as uncovering real events that transpired in the past. As Newall points out, objectivity need not involve "the belief that neutral statements about the world could be made" (Newall, 2009: 175). Instead, the soundness of historical explanation in these terms is related to the extent to which those conditions could justifiably explain events, even if this explanation is supplemented by additional causes that show how it is that these events occurred.

It follows that social conditions are partly akin to Kantian categories. They are necessary conditions for knowledge, in general, and thus we cannot have any historical knowledge independently of them. Yet, unlike Kantian categories, it is not the case that any *particular* social conditions are necessary. Thus, specific social conditions partially constitute only sufficient conditions for particular historical explanations rather than necessary ones. In other words, though a historical explanation can only be offered by relying on the social context, a particular context can only provide a sufficient ground for the emergence of political and social phenomena. This allows us to limit appealing to necessity as much as possible in historical explanation. These social conditions are, on the one hand, contingent as they are the product of a specific period of time and historical circumstances. Moreover, they offer criteria for specifying what renders certain narratives acceptable over others. One consequence of this view is that different historical contexts can produce significantly similar phenomena. Given that historical conditions contribute to establishing the sufficiency of the appearance of a historical event, different historical conditions could also bring about similar events in multiple settings. It is useful to consider an example here. Jack Harrington argues that the conception of citizenship in both Algeria and India shared aspects following from attempting to replace the view of "Imperial citizenship" with a national sense of belonging. This led to problematically introducing a promise of citizenship that is open to all while simultaneously partially preserving already-existing hierarchical divisions in society (Harrington, 2015: 55–56, 63–64). This illustrates a clear example of political problems in a decolonial setting that arose out of particular and different social and historical conditions, where French and British colonialism played a significant role. Yet, in both cases, there is significant overlap in the outcome that allows for a general characterisation of the political problem of citizenship, and thus allows for offering general prescriptions for inclusion when it comes to issues concerning what constitutes national citizenship. A historical analysis of colonial practices in both Algeria and India is, therefore, necessary to form an adequate understanding of the complex relations that led to the exclusion of marginalised groups in society. Providing a historical account in this case also enables us to recognise the intersections in the political concept of citizenship under both models, and thus provides the means to develop critical prescriptions that would apply across both contexts without jettisoning the specificity of the context on the one hand, and while avoiding appealing to universal dictates on the other.

In a different example, one can examine the similarities in the political organisation we find in Lebanon and Iraq that have eventually produced two of the most severe economic crises. The consociational sectarian democracies we find in both countries have emerged out of very different historical conditions. In the Lebanese case, the political setup began during Ottoman times and was fully institutionalised in the early twentieth century following the First World War. In Iraq, a similar political setup came into being as an outcome of a violent ousting of a dictatorship through invasion almost two decades ago. Despite marked differences in the organisation of both states, the political problems resulting from both political economies have been similar in nature leading to mass protests protesting against poor governance. Given the similarities in the political processes that have led to the current crises, it is clear that possible resolutions to the overlapping problems can be general in scope despite the different conditions of their historical emergence.[8]

Several rejoinders can be offered to the view I endorse. First, one can argue that my argument reintroduces the same problems that logical positivists sought to present in terms of sufficiency and necessity. The argument I put forward, however, does not aim to reduce all causation in history to necessary and sufficient conditions. In fact, I am less concerned with the question of what counts as a cause here and more concerned with identifying background conditions that render certain phenomena possible, or at least contribute to the sufficiency of those phenomena. This does not imply that the conditions themselves are enough for the emergence of particular social and political structures. In other words, some conditions are needed for a particular social phenomenon to arise, yet these conditions will often need to be joined with some causal event (or events) that actually produce the complex social relations at hand. Second, logical positivists were interested in resolving the question of the function of historical explanation by reducing causation to sufficient and necessary conditions. My argument does not offer a resolution to the problem of identifying what constitutes an explanation in historiography. Instead, given that my interest is the political use of history, what I show instead is that once we understand historical conditions as contributing to the emergence of social phenomena through partial sufficiency, this allows for moving beyond the universalism/relativism dichotomy. In addition, and unlike the picture positivists paint, I am not committed to the objectivity of historical narratives metaphysically in terms of uncovering a real past.

One main advantage of my account, furthermore, is that it eschews falling into the trap of rejecting objectivity altogether. The relativist line of argument will seek to undermine the need for an objective analysis of history. From a decolonial point of view, the relativist allows for overhauling Eurocentric disciplines in order to make room for local forms of knowledge. Only then would it be possible to shed light on marginalised views. Yet, it is important to note that even the decolonial account will require some commitment to objectivity. As Newall argues, the absence of truth or falsehood in our narratives implies that we have no means of exposing any form of injustice (Newall, 2009: 177). It follows that the same would also apply to the decolonial critique itself. We want to cement the assertion that colonialism has indeed been terrible. Yet, that requires introducing a reliable historical account of the practices of colonialism that shows why it was terrible. That also involves a commitment to the idea that historical accounts that seek to belittle the harmful consequences of colonialism are themselves inaccurate and are not up to standard. For the decolonial critique to be effective, this standard can only be a standard of objectivity, despite it not being universal. In addition, objectivity is also crucial for decolonial critique in another sense. The absence of an objective ground for critique could slip into defending conservatism under the pretext of defending authentic forms of life. We, therefore, also want to avoid the uncritical acceptance of non-Eurocentric practices. Avoiding the trap of a reactive decolonial critique can only involve introducing a standard of objectivity.

Understanding background social conditions as contributing to the emergence of historical events in terms of sufficiency opens the space for general, but not universal, prescriptions in politics. If different phenomena can produce identical problems, then general political prescriptions can be specified in order to work towards social and political change. This allows us to avoid committing to a pernicious relativism that is unable to offer a rational basis for critique. Ultimately, determining how we ought to react to our political condition must result from beginning with a historical analysis. Only then can we identify the overlap between our conditions and the conditions of others. While avoiding universalising political prescriptions without appealing to history, ascribing this role to historical explanation allows us to put forward historically informed normative claims. Finally, recognising that politics must be grounded in historical analyses takes us a step closer towards avoiding a Eurocentric conception of politics, by rejecting decontextualised universal normative demands that overlook the specificity of political problems.

Conclusion

I have argued that the examination of the problem of objective historiography from different standpoints as outlined above allows us to yield several conclusions. First, an objective conception of history is needed for political critique to be possible in general, and for decolonial critique in particular. Second, such a view of historical knowledge can neither be committed to realism with respect to the subject matter of historical analysis nor can it aim at identifying general laws in history. Third, philosophical analysis should be concerned with the real world of experience, i.e., the social world, as opposed to *a priori* categories of understanding. This leads to several consequences. To begin with, the philosophy of history is no longer immune to the conclusions that arise from historical analysis, and the political values endorsed philosophically can be themselves transformed and thus subjected to critique. Furthermore, this allows us to recognise that philosophical accounts themselves depend on contingent values that are social, historical, and thus subject to revision. It follows that political philosophy should examine contingent social conditions that bring about present circumstances while recognising the active role of the subject in constructing these accounts. It is here that we can find the overlap between philosophy and history, allowing for our historical analysis to permeate and provide critical insight into the theoretical world of political philosophy. Finally, historicising the conditions that make historical accounts possible opens the space for critiquing historical knowledge in a manner that allows for further inclusion. Given that decolonial thought is grounded in a historical analysis of colonial practices, critique can only be completed by permeating the conditions that make such an analysis possible, which carves the boundaries of what is acceptable knowledge and what is not.

Notes

1 There is a question as to whether Mignolo's "pluritopic hermeneutics" can be understood without adopting a relativistic framework. Mignolo, for instance, argues that his view entails "diversality" instead of relativism (Mignolo, 1995, 2002: 90). Likewise, Escobar's notion of "pluriversality" arguably offers means to avoid collapsing into an open relativism (Escobar, 2020).

2 This chapter assumes the standpoint of non-ideal theory without arguing for it. The distinction between ideal and non-ideal theory was originally introduced by John Rawls, but is now broadly used with reference to the starting point of political analysis. Ideal theory strives to put forward an account of the best political organisation under ideal conditions, and is therefore not concerned with empirical analysis. Non-ideal theory, in contrast, begins with empirical analysis and only afterwards draws insights about theoretical political conclusions (Simmons, 2010).

3 A problem that arises with beginning with non-ideal practices is how our descriptive analysis (or how practices *are*) could inform us as to how they normatively *should* be. This problem is certainly of relevance, but it will not be the focus of this chapter.

4 It is noteworthy that there is a plethora of views within early twentieth-century logical positivism concerning what it means to unify scientific knowledge under one framework. Krzysztof Brzechczyn offers a rich analysis of these various views, but for the purposes of this chapter, I will avoid going into the details (Brzechczyn, 2009).

5 This claim has been subjected to criticism by Carr and Ricoeur who argue that a phenomenological analysis shows that we experience narratives. As Ankersmit argues, however, one can maintain that stories are told rather than lived, and that applies to us even when we are reflecting on our own narrative, or when we represent our own experiences to ourselves (Ankersmit, 2009: 205; Ricoeur, 2012; Carr, 1986).

6 There is a clear commitment to a Foucauldian approach underlying my view. To put it in Foucault's words, a historical analysis aims at "separat[ing] out, from the contingency that has made us what we are [or from contingent conditions of possibility], the possibility of no longer being, doing, or thinking what we are, do or think" (Foucault, 1998: 315–6).
7 Having a clear and consistent account of objectivity in historical knowledge is a difficult problem that I will not attempt to resolve here. Peter Newall examines various promising accounts of objectivity that follow from engagement within the discipline instead of a commitment to realism about the past (Newall, 2009).
8 Nancy Ezzeddine and Beatrice Noun analyse the similarities and differences between the two political economies shedding light on the role of different factors contributing to the difficult tracks to reform (Ezzeddine & Noun, 2020).

References

Ankersmit, F. (2009). Narrative and Interpretation. In A. Tucker (Ed.), *A Companion to the Philosophy of History and Historiography* (pp. 199–208). Oxford: Blackwell Publishing.

Aristotle (2009). Poetics. In R. McKeon (Ed.), *The Basic Works of Aristotle* (pp. 1455–1487). New York: Random House Publishing Group.

Brzechczyn, K. (2009). Logical Empiricism and Logical Positivism. In A. Tucker (Ed.), *A Companion to the Philosophy of History and Historiography* (pp. 416–426). Oxford: Blackwell Publishing.

Carr, D. (1986). Narrative and the Real World: An Argument for Continuity. *History and Theory*, *25*(2), 117–131.

D'Amico, R. (2009). Historicism. In A. Tucker (Ed.), *A Companion to the Philosophy of History and Historiography* (pp. 243–252). Oxford: Blackwell Publishing.

Danto, A. C. (2007). *Narration and Knowledge*. New York: Columbia University Press.

D'Oro, G. (2009). Historiographic Understanding. In A. Tucker (Ed.), *A Companion to the Philosophy of History and Historiography* (pp. 142–151). Oxford: Blackwell Publishing.

Escobar, A. (2020). *Pluriversal Politics: The Real and the Possible*. Trans. by D. Frye. London and Durham: Duke University Press.

Ezzeddine, N., & Noun, B. (2020). (rep.). *Iraq and Lebanon's tortuous paths to reform*. Retrieved November 29, 2022, from https://www.clingendael.org/sites/default/files/2020-12/iraq-and-lebanons-tortuous-paths-to-reform.pdf

Foucault, M. (1998). What is Enlightenment? In P. Rabinow (Ed.), *Ethics: Subjectivity and Truth. Essential Works of Foucault, 1954–1984, Vol. 1* (pp. 303–320). New York: The New Press.

Foucault, M. (2002). *Archaeology of Knowledge*. London: Psychology Press.

Foucault, M. (2012). *The Order of Things: An Archaeology of Human Sciences* (Reissue edition). New York: Vintage.

Gil, T. (2009). Leopold Ranke. In A. Tucker (Ed.), *A Companion to the Philosophy of History and Historiography* (pp. 383–392). Oxford: Blackwell Publishing.

Harrington, J. (2015). The Imperial Citizen: British India and French Algeria. In E. Isin (Ed.), *Citizenship after Orientalism: Transforming Political Theory* (pp. 53–69). London: Palgrave Macmillan.

Hempel, C. G. (1942). The Function of General Laws in History. *The Journal of Philosophy*, *39*(2), 35–48. doi: 10.2307/2017635

Kosso, P. (2009). Philosophy of Historiography. In A. Tucker (Ed.), *A Companion to the Philosophy of History and Historiography* (pp. 9–25). Oxford: Blackwell Publishing.
Kuhn, T. S. (2012). *The Structure of Scientific Revolutions: 50th Anniversary Edition.* Chicago: University of Chicago Press.
Mandelbaum, M. (1938). *The Problem of Historical Knowledge.* New York: Liveright Publishing Corp.
Mignolo, W. (1995). *The Darker Side of the Renaissance: Literacy, Territoriality, and Colonization.* Ann Arbor: University of Michigan Press.
Mignolo, W. (2002). The Geopolitics of Knowledge and the Colonial Difference. *South Atlantic Quarterly*, *101*(1), 57–96. doi:10.1215/00382876-101-1-57
Newall, P. (2009). Historiographic Objectivity. In A. Tucker (Ed.), *A Companion to the Philosophy of History and Historiography* (pp. 172–180). Oxford: Blackwell Publishing.
Rickert, H. (1962). *Science and History: A Critique of Positivist Epistemology.* Trans. by G. Reisman. New Jersey: Princeton University Press.
Ricoeur, P. (2012). *Time and Narrative, Volume 1.* Trans. by K. McLaughlin & D. Pellauer. Chicago: University of Chicago Press.
Said, E. W. (2004). *Orientalism.* New York: Vintage.
Simmons, A. J. (2010). Ideal and Nonideal Theory. *Philosophy & Public Affairs*, *38*(1), 5–36.
Strauss, L. (1949). Political Philosophy and History. *Journal of the History of Ideas*, *10*(1), 30–50. doi: 10.2307/2707198
Tucker, A. (2009). Causation in Historiography. In A. Tucker (Ed.), *A Companion to the Philosophy of History and Historiography* (pp. 98–108). Oxford: Blackwell Publishing.
White, H. (1975). *Metahistory: The Historical Imagination in Nineteenth-Century Europe.* Baltimore: Johns Hopkins University Press.
White, H. (2010). *The Fiction of Narrative: Essays on History, Literature, and Theory, 1957–2007.* Baltimore: Johns Hopkins University Press.
Zammito, J. (2009). Historians and Philosophy of Historiography. In A. Tucker (Ed.), *A Companion to the Philosophy of History and Historiography* (pp. 63–83). Oxford: Blackwell Publishing.

2 The Recalcitrance of White Ignorance

Laurencia Sáenz Benavides

Introduction

As an ideology, whiteness typically conceals the mechanisms through which race-based oppression is reproduced. One aspect of this ideological formation is visible in the white imaginary, around which some of the desires, emotions, identifications, and affiliations of white subjects organise. Under the heading of "white ignorance", Charles Mills has accounted for a pattern of cognition that is typical, although not exclusive of, white people. In this sense, white ignorance designates a particular effect of whiteness as an ideology in the formation of white subjectivity: the distinctive cognitive limitations of whites. Mills' notion of white ignorance is especially relevant for reflecting on decolonising political concepts since white ignorance is one of the processes through which the coloniality of political concepts has been produced.

I will proceed in the following way. In section "White Ignorance and the Causal Role of Race", I will flesh out the five cognitive processes that, according to Mills, are responsible for the production and reproduction of white ignorance, namely, *conception, perception, memory, testimony*, and *motivational group interest*. Mills' framework gives us key insights as to why decolonising political concepts is not an easy, straightforward task. However, I will argue that to better account for the *recalcitrant* persistence of white ignorance, motivational group interest should be seen as its driving force and, as such, it should be thought of as having a different status with respect to the four other cognitive processes. In section "Affective Aspects of White Ignorance", I will develop this argument by showing how the conative and affective dimensions of motivational group interest can help us explain why white ignorance is so difficult to eradicate. Drawing on the works of Audre Lorde, Toni Morrison, George Yancy, and Sara Ahmed, I aim to show how white ignorance is rooted in affective strategies, through which whites maintain their power. In this sense, given the importance of these affective elements, any attempt to decolonise political concepts will need to contend with the conative force that holds their matrix in place.

DOI: 10.4324/9781003293460-4

White Ignorance and the Causal Role of Race

According to Mills, white ignorance emerged during the modern period, playing a key instrumental role in the European colonial project. It is not any kind of ignorance that people who happen to be white are prone to, but one in which race, understood as a social construct, "plays a crucial causal role" (Mills, 2007: 20). It designates a specific kind of "group-based cognitive handicap" (Mills, 2007: 19). Just as whiteness, as an ideology, influences the views of people who are not white, white ignorance is not *exclusive* to white people; it can affect non-white people as well. Non-whites may exhibit forms of white ignorance when, for example, adhering to claims that in post-racial, colour-blind, and meritocratic Western societies, race no longer has any social significance. However, whites as a group will be more typically prone to it.[1] In this sense, Mills points out, the causal role of race in white ignorance is social-structural, rather than physio-biological. As Linda Martín Alcoff has analysed, Mills' key contribution lies in providing a structural argument, according to which "whites have a *positive* interest in 'seeing the world wrongly', to paraphrase Mills" (Martín Alcoff, 2007: 47).

White ignorance is structural and *structuring*. While in Mills' analysis, ignorance comprises both an absence of true beliefs and the presence of false beliefs, white ignorance is best thought of as a *process*, rather than as a product. It is not to be understood in a passive sense, as a mere lack of knowledge. It is an ignor*ing* (Spelman, 2007); an active process of miscognition that lodges itself in the very processes of knowledge production, "[presenting] itself unblushingly as knowledge" (Mills, 2007: 13). The very processes through which white ignorance structures our ways of un-knowing are made invisible: "Racial seeing as such is not open to view; the processes of racialization that come to structure our social perceptions are not seen, and yet our perceptual habits and our field of vision cannot escape them" (Medina, 2013: 54). For example, the concept of colour-blindness structures our ways of (un)knowing, (mis)perceiving and (un) feeling rather as though we were looking at a fishbowl, as Toni Morrison's image illustrates:

> The glide and flick of the golden scales, the green tip, the bolt of white careening back from the gills; the castles at the bottom, surrounded by pebbles and tiny, intricate fronds of green; the barely disturbed water, the fleck of waste and food, the tranquil bubbles traveling to the surface.

Is made possible by the bowl, "the structure that transparently (and invisibly) permits the ordered life it contains to exist in the larger world" (Morrison, 1992: 17). In this sense, white ignorance is a meta-ignorance, an ignorance that is unaware of itself. This aspect partly explains its pervasiveness: it is difficult to challenge something that typically conceals itself. It matters therefore to *see the bowl*, i.e., to make this kind of racial miscognition *visible*, to unmask its hidden mechanisms, in order to fight against its reproduction.

Mills identifies five key cognitive processes involved in white ignorance: *perception, conception, memory, testimony, and motivational group interest*. Identifying and analysing how these processes operate may allow us to fight against their reproduction. However, as I will contend, motivational group interest stands apart from the other four, insofar as it is not merely a cognitive, but also a *conative* process. It is aimed at the fulfilment of an action. In this sense, it seems to play a distinctive and more decisive role in the pervasive and self-concealing aspect of this group-based cognitive handicap.

Conception and Perception

The five processes analysed by Mills are analytic category distinctions for processes that are in practice always intertwined:

> Perception is also in part conception, the viewing of the world through a particular conceptual grid. Inference from perception involves the overt or tacit appeal to memory, which will be not merely individual but social. As such, it will be founded in testimony and ultimately on the perceptions and conceptions of others.
>
> (Mills, 2007: 24)

The role of conception and perception can be seen in the way European imperialist expansion during the modern period entailed forging and mobilising specific concepts that orientated how the colonised world was perceived. This process involved the emergence of a particular epistemic principle, "white normativity", which asserted the moral, aesthetic, cultural, and intellectual superiority of the European world. From this normative principle stemmed a series of distorting epistemic practices and myths that misrepresented and/or suppressed facts about the colonised peoples. For example, the concept of the 'savage' played an instrumental role in justifying European imperialism. As such, the concept is inherently tied to the colonial project, and orientated the representations that whites formed of non-whites, as well as colonial practices:

> Even a cognizer with no antipathy or prejudice toward Native Americans [would] be cognitively disabled trying to establish truths about them insofar as such a category and its associated presuppositions [would] tend to force his conclusions in a certain direction, will limit what he can objectively see.
>
> (Mills, 2017: 62)

White ignorance, therefore, encompasses both explicit and implicit racist beliefs and attitudes. As a social-structural phenomenon, it does not always rest on ill intent. It is as though, by the inertial force of inadequate categories and perceptions, cognitive dysfunctions became part of the habits of whites even when they consciously reject racism as a form of injustice.

In a context of *de jure* white supremacy, such as under Jim Crow legislation in the United States or the *apartheid* regime in South Africa, the "racialized causality" in white ignorance tends to be more direct, and people's general beliefs tend to be more explicitly racist. By contrast, in a context of *de facto* white supremacy – more characteristic of contemporary liberal democracies in which explicit racist beliefs are socially condemned and explicit racist laws have been abolished – this racialised causality still operates indirectly, even in individuals who do not consciously and explicitly endorse racist views. The kind of white ignorance that is pervasive in a *de facto* white supremacist context is exemplified by 'colour-blindness': the idea that we live in a world in which race is no longer relevant and in which everyone is equal before the Law, regardless of skin colour. In the context of colour-blindness, white ignorance is manifest in interpersonal relations through utterances such as "When I look at you, I do not see color" (Medina, 2013: 40). Often well-intentioned, these rejections of racism ignore the ways that race continues to play an important role in the lives of those who suffer racial discrimination. As José Medina points out, "the complete refusal to see color in a racist society involves implicitly the refusal to acknowledge the force of racist prejudices and their insidious impact on interpersonal dynamics" (Medina, 2013: 40). The belief that Western liberal democracies are colour-blind reduces white people's ability to perceive social-structural injustice for what it is. For many, if we all have equal opportunities, and if socio-economic 'success' is a matter of individual perseverance and personal hard work, then those who 'fail' by those standards must do so because they lack the motivation and the right set of values. White ignorance encourages a tendency to blame the groups who suffer racial discrimination for their problems. Its cognitive distortions have an impact on people's moral sensitivity.

Memory and Testimony

White ignorance involves forms of collective amnesia. For example, many North, Central, and South American countries still commemorate October 12[2] as 'Columbus Day', or celebrate it as the 'encounter of cultures', thereby underplaying the genocidal violence of colonisation. In Spain, the same date commemorates the National Holiday of Spain (Fiesta Nacional de España) known as 'Día de Hispanidad' (Hispanity Day). On the official Tourism webpage of the city of Madrid, the holiday is described as celebrating "the cultural and linguistic expansion of the country beyond the European continent", and includes a military parade in honour of the King and Queen of Spain, honouring "those who sacrificed their life for Spain" (my translation).[3] No mention is made however of the native populations exterminated by genocide or to those who still suffer the harmful impacts of colonialism, such as marginalisation, poverty, and cultural and linguistic extinction. Such distortions of memory produce white ignorance in a form that makes it difficult for most white people to recognise their identities as indissolubly tied to the histories of oppression that have systematically advantaged them. In this sense, collective amnesia perpetuates the coloniality of political concepts such as national and cultural identity.

The cognitive processes of memory involved in white ignorance, as a practice that has distorting effects on social cognition, leads to having inadequate knowledge of the lived realities of non-whites, as well as of white people's own historical and current position of privilege. This entails a form of 'moral ignorance' that limits white people's capacity to understand the moral wrongs of oppression, and their participation in it: "Whites(…) experience genuine cognitive difficulties in recognizing certain behaviour patterns as racist, so that (…) they will be morally handicapped simply from the conceptual point of view in seeing and doing the right thing" (Mills, 1997, as cited in Sullivan, 2015: 128).

Collective amnesia and other distortions of memory, perception, and conception have an impact on white people's receptivity to the *testimony* of those oppressed by white supremacy. For example, white people will have a greater tendency to dismiss racially oppressed people's accounts of the injustices they suffer, either by not believing them or attributing their perceptions to being oversensitive, paranoid, or manipulative ('using the race card').

A Recalcitrant Ignorance

The previous are some of the ways in which conception, perception, memory, and testimony generate and perpetuate white ignorance. As we saw above, white ignorance is not a mere lack of information but has an active dimension, and these four aspects show how it is active insofar as it is *productive* of patterns of miscognition that will typically cognitively handicap white people. These processes being closely intertwined, decolonising political concepts will require modifying our patterns of conceiving, perceiving, remembering, and listening. However, white ignorance is active in another sense, which goes beyond the productive character of the four processes discussed so far. Mills describes it as "an ignorance that resists, (…) an ignorance that fights back (…) an ignorance militant, aggressive, not to be intimidated, an ignorance that stays active, dynamic, that refuses to go quietly (…) presenting itself unblushingly as knowledge" (Mills, 2007: 13). José Medina also highlights this active dimension of white ignorance: "It is a recalcitrant ignorance, hard to eradicate, that is rooted in active patterns of cognitive interaction and in habitual ways of perceiving, listening, talking, thinking, and acting" (Medina, 2013: 39).

How should we account for the "recalcitrant", "militant", and "aggressive" character of white ignorance? While the four cognitive processes previously discussed do account for some of the ways in which white ignorance is active and pervasive, they seem insufficient in themselves to explain its *recalcitrance*, as Mills and Medina describe it in the previous excerpts. Conception, memory, testimony, and perception certainly produce distorted beliefs and inadequate knowledge, but such distortions could be corrected relatively straightforwardly by putting in place mechanisms for cognitive transformation. Mills does believe in the necessity of such transformations at the cognitive level. As Alison Bailey has analysed, undoing white ignorance requires "a historical revisionist project and a program of cognitive reform":

> If white ignorance is the product of an "inverted epistemology," then a revisionist history buttressed by a race-sensitive program in cognitive science should replace race-ignorant with race-cognizant knowing. The historical project speaks for itself: read history through a racial lens, and get the bigger picture.
>
> (Bailey 2007: 81)

On the other hand, cognitive reform is needed in order to dismantle the "racialized moral psychology" that operates in white ignorance; according to Mills, cognitive science can provide instruments of analysis and techniques for countering the "dysfunctional thought patterns" (Bailey, 2007: 81).

However, Bailey contends that the image of an "inverted epistemology" is unhelpful, as it "suggests that solutions lie in *reinverting* the epistemology, as one would turn a sweater right side out" (Bailey, 2007: 81). This has the risk of reinscribing a "logic of purity" (Lugones, 2003), a way of thinking that underpins white fantasies of unity and wholeness, and whose effects are to "erase, control, and distort the true multiplicity of all social beings" (Bailey, 2007: 83). This logic underpins social contract theory as well. Bailey defends the advantages of Lugones' "logic of curdling", "a pluralistic logic that recognizes all persons as complex multiple subjects" (Bailey, 2007: 84). In following Lugones' logic of *mestizaje*, Bailey argues, we are in a better measure to undo white ignorance, and therefore, we may add, in a better position to decolonise the political concepts that may be informed by white ignorance:

> If privileged groups' desire for wholeness gives rise to the split-separation thinking that teaches white folks to see the world wrongly, then combatting white ignorance will require that white folks abandon the myth of unified wholeness and learn to see our *own* multiplicity.
>
> (Bailey, 2007: 91)

Whites must learn to think of themselves as "curdled beings", reckoning with the ambiguity, impurity, and messiness that are part of all human experiences.

In line with Bailey's and Lugones' argument, I contend that thinking of ourselves as multiplicitous, "curdled" beings, requires going beyond white ignorance's cognitive dimensions, by exploring its affective roots. In this sense, it remains unclear how Mills' project of historical revisionism and cognitive reform could address the "militant" and "aggressive" character of white ignorance, insofar as such characterisation suggests that there is not merely a cognitive dysfunction, but also that white ignorance has a *conative* dimension, for which transformations at the cognitive level seem insufficient.

Motivational Group Interest

This brings us to the fifth cognitive process, namely, *motivational group interest*. While perception and conception in white ignorance consist in patterns of unseeing and misconceiving, motivational group interest involves more specifically a volitional dimension, such as a will to misperceive, misremember, misrepresent, etc. How should we think of this volitional element if, on the one hand, Mills' structural analysis has established that white ignorance does not necessarily rest on ill intent? On the other hand, how should we think of motivation in the case of this group-based cognitive dysfunction without falling into individualistic analyses that explain racism in terms of the bad intentions of individual agents (Garcia, 1996)?

Let us go back to Mills' description of this motivational element: "At all levels, interests *may* shape cognition, influencing what and how we see, what we and society choose to remember, whose testimony is solicited and whose is not, and which facts and framework are sought out and accepted" (Mills, 2007: 24. My italics). The conditional form ("may") in the previous passage suggests that motivational group interest is not a necessary component of white ignorance; there may be instances of white ignorance that lack such motivational component. However, Mills does consider it to be a "central causal factor in generating and sustaining white ignorance" (Mills, 2007: 34). Even if it is not a necessary element, this does not exclude its centrality. One way to think of it is to see motivational group interest as a driving force at the group level but to a lesser degree at the individual one. In this sense, if in general whites "see black interests as opposed to their own" (Mills, 2007: 35), this perception may not be a motivating factor for all individuals. As a phenomenon that concerns groups, it is best thought of as a general tendency, to which individuals may be more or less prone to, depending on different factors, such as education, cultural environment, relationships, personal sensitivity, etc.

Noting that an analysis of the links between group interests and cognition is lacking in the scholarship of social epistemology, Mills' argument suggests that taking motivational interests into consideration need not lead to individualistic explanations. Whilst in the Marxist tradition, it has been broadly accepted that "if exploitative socio-economic relations are indeed foundational to the social order, then this is likely to have a fundamental shaping effect on social ideation" (Mills, 2007: 34), the same kind of phenomenon needs to be recognised with respect to matters of race. Mills argues that whites' perception of their own group interests as threatened by black interests drives their preferences and shapes their cognitive practices. Therefore, motivational group interest plays a key role in the recalcitrant aspect of white ignorance, along with the other elements identified by Mills, and it manifests itself in "the *refusal* to perceive systemic domination, the *convenient* amnesia about the past and its legacy in the present, and the *hostility* to black testimony" (Mills, 2007: 35. My italics).

The question of ill intent and motivation can be solved if we think of motivation in a broad sense. In many cases, motivation drives people's cognitive habits without it being reducible to a matter of ill intent. For example, our affects and emotions can often drive our curiosity or lack thereof without us being fully aware of it. However, Mill's cognitive framework does not allow us to further make sense of the conative aspects of white ignorance. Although Mills takes motivational group interest to be a "central causal factor in generating and sustaining white ignorance", this fifth conative/cognitive component occupies a marginal place in his analysis. But if this is a key driving force behind the cognitive dysfunctions of white ignorance, then any attempt to undo it – any effort to decolonise white ignorance as a generator of the coloniality of political concepts – will have to address the conative aspect as well. If we are to understand how motivational group interest plays a key role in the recalcitrance of white ignorance, we, therefore, need to explore its affective roots. The recalcitrant aspect of white ignorance, as rooted in conative/affective elements, suggests that any attempt to combat it will have to go beyond historical revisionism and cognitive reform: undoing white ignorance will require tackling the affective mechanisms that underpin motivational group interest.

Affective Aspects of White Ignorance

White ignorance is one of the forces that sustain white privilege insofar as it allows the privilege to remain unacknowledged, and thus unchallenged, by those who benefit from it. As we have seen thus far, although Mills attributes to motivational group interest a key role in producing and sustaining white ignorance, his analysis is primarily concerned with the *cognitive* dimensions of this phenomenon. However, as José Medina and Shannon Sullivan argue, white ignorance does not simply operate at the level of *beliefs*, which has been the main focus of Mills' analysis. White ignorance is embedded in the emotional habits that characterise whiteness as an embodied experience: "The racial meanings inscribed in the body become part of the underlying structure of our perceptual habits, that is, part of the taken-for-granted background against which our social perceptions take place" (Medina, 2013: 54–55). In this sense, they are not only explained by the suppression or distortion of facts, by forms of collective amnesia and systematic "epistemic oppression" (Dotson, 2012).

I will now turn to examine some of the ways in which white ignorance not only entails forms of unknowing but also of *unfeeling* (Medina, 2013). I will then highlight how its embodied dimensions are connected to patterns of emotions such as disgust, contempt, and fear. From these analyses, we can identify a series of affective patterns that constitute the emotional sources from which white ignorance extracts its motivational force. It is these affective aspects, I contend, that significantly contribute to its pervasiveness. Any effort to decolonise political concepts will therefore need to contend with the affective elements that sustain white ignorance as a mechanism through which the coloniality of political concepts is reproduced.

Affective Numbness

José Medina characterises the concept of colour-blindness as carrying a particular kind of *insensitivity* to matters of race, which is both cognitive and affective:

> racial insensitivity may involve the failure to see the social *relevance* of race in one's interactions, and this failure is not simply a cognitive deficit, but an affective failure: it involves the inability to *feel* concerned and to have an entire array of *emotions* such as empathy, sympathy, compassion, etc. This is why those who do not see the social relevance of racial aspects of social experience often charge those who do as being *oversensitive*, as having a thin skin or feeling too much when racial elements are present in social interactions.
>
> (Medina, 2013: 49)

Medina identifies different kinds of affective numbness linked to racial insensitivity. A first consists in feeling indifferent or apathetic in relation to a particular social group. This may be linked to other phenomena in the cognitive realm that Medina identifies as "epistemic laziness", a group-based lack of curiosity (Medina, 2013: 33) that produces narrow-mindedness and arrogant patterns of behaviour. It is the indifference or apathy characteristic of racist ethnocentrism, which María Lugones (2003) also links to laziness and arrogance: "(…) the disrespectful, lazy, arrogant indifference to other cultures that devalues them through not seeing appreciatively *any* culture or cultural ways except one's own when one could do otherwise" (Lugones, 2003: 44).

A second kind of affective numbness identified by Medina consists not in being indifferent, but rather in feeling concerned by issues of racial injustice in the abstract, and not knowing how to engage with them, i.e., being *affectively blocked* in one's concrete responses to racial injustice. Such affective blockages have been described by Lugones as "infantilization of judgement" – the tendency of white people to take refuge in their good, innocent intentions when challenged about their participation in racism, and their general inability to take responsibility for their actions when these display signs of ethnocentrism and racism:

> They have turned into children, incapable of judgement, avoiding all commitment except against racism in the abstract, paralyzed as responsible beings, afraid of hostility and hostile in their fear, wedded to their ignorance and arrogant in their guilty purity of heart.
>
> (Lugones, 2003: 48–49)

In Lugones' account, the emotional element is salient: a paralysing fear of being challenged in one's racism, an attachment to forms of ignorance as a way to deflect responsibility for a particular behaviour, and the arrogance implied in claiming that, if one's intentions are good and 'pure', one cannot be challenged for one's participation in racism.

In Medina's and Lugones' account, white ignorance consists in a lack of sensitivity to racial injustices that affect non-whites, as well as a reduced self-awareness of one's involvement in such issues. Medina's discussion of the simultaneously cognitive and affective aspects of white ignorance highlights the emotional deficiencies that privileged subjects may display. White ignorance, in the form of a reduced sensitivity, such as an atrophied emotional capacity for empathy, translates into a diminished moral capacity. Insensitive subjects lack the tools for properly understanding how their own behaviour, perceptions, and ways of interacting perpetuate forms of racial harm. In this way, the cognitive and emotional limitations of privileged subjects may translate into moral failings. Therefore, part of addressing racial injustice requires attending to the ways in which "cognitive and affective structures work together, or fail to work together" (Medina, 2013: 50).

Following Fanon (2008), Medina notes that many aspects of the cognitive and affective insensitivities displayed by white subjects are closely linked to the particular ways in which, in Western cultures, white is the homogenous, invisible norm, "the color of the unmarked mainstream subject" (Medina, 2013: 50). In the white imagination, whiteness is experienced by white subjects as a neutral absence of colour. Furthermore, the mechanisms through which racialised perception is produced also tend to be invisible. Therefore, when white persons see themselves as completely free from racial prejudice, and think of themselves as *unraced*, it is not only that many socially relevant aspects of race tend to remain unseen/unfelt in the ways that Medina describes. More than that, the very processes that structure such ways of seeing (or unseeing) are made invisible: "Racial seeing as such is not open to view; the processes of racialization that come to structure our social perceptions are not seen, and yet our perceptual habits and our field of vision cannot escape them" (Medina, 2013: 54).

White Ignorance as an Embodied Unconscious Habit

White ignorance does not only consist in cognitive habits, Shannon Sullivan contends. It also lies in a series of embodied unconscious habits. In this way, as an embodied habit, white ignorance shapes and helps sustain white privilege by concealing it from those who reap its benefits. The notion of habit

> helps explain how white privilege often functions as if invisible. Habits are the things that we do and say 'without thinking.' They are the mental and physical patterns of engagement with the world that operate without conscious attention or reflection.
>
> (Sullivan, 2006: 4)

Habits are "environmentally constituted"; they are ways of "transacting with the world" that become constitutive of the self (Sullivan, 2006: 2).

In what sense are habits formed in transaction with the environment? The body and the psyche are constituted by the environment and, simultaneously, productive of the environment.

In order to understand the transactional aspect of habit, take the image of a pair of old, used shoes. If I wear the same pair of shoes for years for my daily stroll in the forest, the cadence of my step will be, with the passing of time and use, reflected in their shape. The constant contact with the path of the forest will smooth their soles, bend their tips, and scrape their heels in a particular way. But my steps – along with all of those who take the same path – will also leave their trace, and contribute to giving the path its characteristic shape. My body as a whole may be modified by this daily stroll as well, including by the shoes I wear. If, as it turns out, the shoes were not adapted to walking in the forest, I may end up with chronic tendinitis or back pain. My old shoes were *constituted by* the environment in which they were used (my feet, my steps, the forest path), but they were also *constitutive of* the environment (they gave me back pain; they helped smooth the path of the forest; etc.).

Habits can be modified. However, their unconscious[4] dimension is one of the factors that help explain their *resistance* to change: "They actively thwart the process of conscious reflection on them, which allows them to seem non-existent even as they continue to function" (Sullivan, 2006: 5–6). Sullivan offers the case of an under-confident female student in predominantly male academic settings; she will tend to "present her views in class apologetically"; she will tend to "contract her body inward as she sits": "[T]he student in my example has an inhibited style of engaging with the academic world that is inseparably psychical and bodily" (Sullivan, 2006: 24).

Habits of white privilege will naturally differ depending on whether the context is marked by more conscious and deliberate forms of racism, and more unconscious and implicit manifestations of this phenomenon. Sullivan's case study focuses on the United States. In her view, the transformation from a *de jure* ("Jim Crow") to a *de facto* racism (in twenty-first-century United States) goes with a transformation of the ways in which white domination operates. In a *de jure* white supremacist context, the patterns of domination are more conscious, explicit, visible, and deliberate. In a *de facto* racist context, the forms of white domination tend to be more invisible, implicit, and unconscious. Whilst white domination is always a mixture of white supremacy and white privilege, white domination's current *modus operandi* is a combination in which white privilege is present in higher proportions (Sullivan, 2006: 5).

A salient characteristic of white domination in the context of today's colour-blindness is that it operates while, and all the more effectively *because*, it makes itself invisible. This is why unconscious habits of white privilege need to be brought to visibility, analysed, and challenged, so as to be potentially disrupted or modified. If, as Sullivan argues, in the context of white supremacy, racism is consciously and explicitly embraced by most white people (supported institutionally, legally, economically, and culturally), one may expect that white people, in general, will display patent patterns of emotions that, by the moral standards of today's allegedly 'colour-blind' US society, would be deemed unacceptable, such as pleasure taken in acts of humiliation and cruelty, as well as overt patterns of contempt, disgust, and hatred. Such patterns of behaviour are described in slave narratives, for example, in Frederick Douglass' description of an act of torture as "the blood-stained gate, the entrance to the hell of slavery, through which [he] was about to pass" (Douglass, 2009: 19). Feelings and expressions of disgust and contempt were common currency in a *de jure* white supremacist context, and the psychosomatic habits of white people were importantly shaped by it. Such expressions of disgust were not exceptional, isolated incidents. As George Yancy notes in reference to Kristina DuRocher's (2011) book *Raising Racists: The Socialization of White Children in the Jim Crow South*:

> DuRocher notes that Alice Harris Kester, the wife of Howard Kester, a prominent white preacher who was influenced by the Social Gospel movement, "confronted one of the southern 'sins', at a Negro Baptist Publishing House lunch. She tried to eat at the same table as African Americans, but could not keep her food down, running home in tears." Both white women appear to be sincere in their efforts at political activism. Yet, their bodies responded in ways contrary to their intentions.
>
> (Yancy, 2017: 245–6)

If the environment plays an important role in the way habits are constituted, and if modifications in the environment produce transformations in habits, one might think that in today's alleged colour-blind regime, such psychosomatic phenomena would have disappeared. However, Sullivan's analysis shows how some of these habits, rather than disappearing, have become *unconscious* in the passage from a white supremacist context to one dominated by white privilege. Something of the old habits remains, despite modification. As mentioned above, it is not that white domination has ceased, but that it has transformed its modes of operation.[5] Many of the psychosomatic habits of white privilege have become "woven into the fabric of [white people's] unconscious" (Yancy, 2017: 34). The legacy of the 'old-fashioned' Jim Crow era racism still acts in white people's bodies.[6]

Beyond Beliefs

Habits of white privilege do not merely take their shape from processes of internalisation of particular beliefs of the propositional and representational kind. Rather, unconscious racial habits are produced and transmitted, as it were, *through the body*. Embodied racial habits are transmitted through bodily actions and reactions in ways that manifest and perpetuate a racial and racist imagination. Drawing on Jean Laplanche's (1989) theory of seduction, Sullivan contends that the unconscious is "initially and continually formed in relationship with concrete others in a sociopolitical world" (Sullivan, 2006: 64).

Laplanche's theory explains the role of adult's seduction[7] in the constitution of the infant's unconscious:

> By means of bodily expressions such as gestures or grimaces – and also, though rarely for babies, by means of spoken words – the adult implants a message about sexuality in the child's body, at least a portion of which the

> child cannot comprehend. The child tries to understand the message, and indeed sometimes succeeds in part. The parts that she does not understand are repressed. These remainders of the attempted translation of the message form the child's unconscious. The etymology of the verb "to seduce" (*séduire*) helps indicate why the process is seductive: in seduction, an adult draws an infant into the adult world in an irresistible fashion, captivating the child in way that he or she does not know how to respond to.
>
> (Sullivan, 2006: 64–65)

Sullivan adapts this view in order to show how children's unconscious is shaped by the adult world through bodily signals, using a passage from Morrison's (2016) *The Bluest Eye*:

> Toni Morrison's novel *The Bluest Eye* provides a helpful illustration of the process of seduction. Morrison demonstrates how the narrator of the novel, a nine-year-old black girl named Claudia, and her older sister, Frieda, are tuned into the adult world around them, receiving its messages even though they do not fully understand them:
>
> Frieda and I are washing Mason jars. We do not hear their [the adults in the other room] words, but with grown-ups we listen to and watch out for their voices… The edge, the curl, the thrust of their emotions is always clear to Frieda and me. We do not, cannot, know the meanings of all their words, for we are nine and ten years old. So we watch their faces, their hands, their feet, and listen for truth in timbre.
>
> Morrison reveals an adult world full of unintended bodily gestures and tones that communicates a great deal of enigmatic meaning to the children in it. From the sound of parents' and neighbors' voices, Claudia and Frieda know that something is up, but they do not fully understand the edgy mood that filters from the living room into the kitchen. The incomprehensible portions of the adults' message – which, in this case, involve the yearning and later angry revulsion generated by a newly arrived boarder in Claudia's home – will become part of each girl's unconscious.
>
> (Sullivan, 2006: 65)

Sullivan identifies a particular feature of her own unconscious habits of white privilege in the association of the smell of cumin "with the (perceived) body odor of Mexicans" (Sullivan, 2006: 68). She traces the origin of this habit in her grandmother's particularly contemptuous tone of voice: "one of the enigmatic messages sent to me regarding race likely originated in the distasteful hiss of my grandmother's voice as she pronounced the word 'Mexican'" (Sullivan, 2006: 69). Of course, Sullivan's particular racist association between a spice, a perceived body odour and an ethnicity, was most likely not simply transmitted by her grandmother. Unconscious habits are formed, she argues, through a "transgenerational crowd" (Sullivan, 2006: 69). Unconscious habits of white privilege, both bodily and psychic, are *recalcitrant* insofar as they resist conscious beliefs or knowledge that could otherwise counter and "correct" the racist associations that they carry.

Recalcitrant Habits and White Narcissism

The recalcitrance of people's psychosomatic habits, Sullivan suggests, is partly explained by the protective function that the latter play in securing white people's sense of self. The invisibility of white privilege to those who benefit from it implies thinking of oneself as race-free, colour-free, but also smell-free while imagining the other (in her example, the Mexicans) as *coloured, raced*, having *an accent*, having *a smell*, etc. If to be *smelly* is to be *dirty*, and dirtiness is associated with moral deficiencies, then, by the same token, to be without smell is to be clean, to be clean is to be pure, virtuous, etc. In some of the examples discussed above, psychosomatic habits of white privilege relate to tasting/digesting, smelling, and hearing. The latter are most probably linked with patterns of *disgust* which may have a protective role in securing whites' sense of self as 'clean', 'pure', etc. We find here further evidence of the active, recalcitrant aspect of white ignorance that Mills partly explained by the existence of "motivational white group interests".

As we can now see, white group interests are secured not only cognitively, via patterns of thought and beliefs but also through affective and bodily habits. Whilst Mills focuses on the cognitive aspects of privilege, Medina and Lugones identify affective blockages that partly account for the persistence of white ignorance. Forms of insensitivity to matters of race prevent whites from acknowledging the ways in which they may be complicit in racial oppression, which in turn allows prejudicial behaviour to remain unchallenged. Sullivan's exploration of the unconscious, physiological habits of whiteness, offers additional elements that explain the affective roots of motivational group interest, which allows us to account for the persistent character of white ignorance.

The Emotions of Oppressors as Seen by the Oppressed

However, the affective and emotional patterns connected to the white embodied self also have a distinctive phenomenology from the perspective of subjects who are targeted by racism, as narratives and phenomenological descriptions of oppression, by the oppressed themselves, show. For example, Audre Lorde's account of one of her first encounters with racial hatred, as a child, provides insight into the kind of emotional patterns of oppressors, which underpin what Mills characterised as motivational group interest.

> The AA subway train to Harlem. I clutch my mother's sleeve, her arms full of shopping bags, christmas-heavy. The wet smell of winter clothes, the train's lurching. My mother spots an almost seat, pushes my little snowsuited body down. On one side of me is a man reading a paper. On the other, a woman in a fur hat staring at me. Her mouth twitches as she

> stares and then her gaze drops down, pulling mine with it. Her leather-gloved hand plucks at the line where my new blue snowpants and her sleek fur coat meet. She jerks her coat closer to her. I look. I do not see whatever terrible thing she is seeing on the seat between us – probably a roach. But she has communicated her horror to me. It must be something very bad from the way she's looking, so I pull my snowsuit closer to me away from it, too. When I look up the woman is still staring at me, her nose holes and her eyes huge. And suddenly I realize there is nothing crawling up the seat between us; it is me she doesn't want her coat to touch. The fur brushed past my face as she stands with a shudder and holds on to strap in the speeding train. Born and bred a New York City child, I quickly slide over to make room for my mother to sit down. No word has been spoken. I'm afraid to say anything to my mother because I don't know what I've done. I look at the sides of my snowpants, secretly. Is there something on them? Something's going on here I do not understand, but I will never forget it. Her eyes. The flared nostrils. The hate.
>
> (Lorde, 2017: 135–36)

Lorde's narrative makes visible the patterns of hatred and disgust that the white woman embodies: her mouth-twitching, her wide-eyed gazing, and her flared nostrils. The white woman's horrified expression signifies to the young Audre that there must be something truly disgusting to justify her reaction ("I do not see whatever terrible thing she is seeing on the seat between us – probably a roach"). The way the white woman's body moves away from her ("The fur brushes past my face as she stands with a shudder") communicates horror, hatred, and disgust in non-verbal ways ("No word has been spoken"). Yet all this is felt by the child, even if she does not fully understand the meaning: "Something's going on here I do not fully understand, but I will never forget it. Her eyes. The flared nostrils. The hate" (Lorde, 2017: 147–48). Lorde's narrative not only shows how racist abuse harms the oppressed, it also sheds light on how racism is connected with particular embodied and emotional configurations in the oppressors.

Furthermore, Sara Ahmed's (2007) phenomenological analysis of whiteness, as viewed from the perspectives of the oppressed, allows us to highlight the emotional aspects of white ignorance. Ahmed takes Fanon's description of what he would have to do if he wanted to smoke next to a white man as the starting point for her analysis of whiteness, which appears affectively marked by comfort for white bodies:

> And then the occasion arose when I had to meet the white man's eyes. An unfamiliar weight burdened me. The real world challenged my claims. In the white world the man of color encounters difficulties in the development of his bodily schema. Consciousness of the body is solely a negating activity. It is a third-person consciousness. The body is surrounded by an atmosphere of certain uncertainty. I know that if I want to smoke, I shall have to reach out my right arm and take the pack of cigarettes lying

> at the other end of the table. The matches, however, are in the drawer on the left, and I shall have to lean back slightly. And all these movements are made not out of habit, but out of implicit knowledge.
>
> (Fanon, 1986: 110–11, as cited in Ahmed, 2007:152)

Fanon captures what it means for him to move in a spatial setting that makes things familiar and easily available to a white body, but not to a Black man. By merely thinking of the movements his body would have to make in order to smoke, he is "burdened" by an "unfamiliar weight". His body becomes "surrounded by an atmosphere of certain uncertainty" under the gaze of the white man. Whiteness, then, structures what bodies can and cannot do, or the easiness with which they can do what they intend to do: "Fanon's example shows the body *before* it is racialised, or made black by becoming the object of the hostile white gaze. (…) If the world is made white, then the body-at-home is one that can inhabit whiteness" (Ahmed, 2007: 153).

Ahmed analyses whiteness as *habitual*, describing how it allows white bodies to be *unproblematic* and *unnoticeable*, while marking non-white bodies as noticeable, unable to 'fit'. Whiteness, as habitual, orientates around its orbit what bodies can and cannot do. If whiteness is what allows white bodies to go unnoticed in a white world, it is what restricts, by the same token, the privilege of being 'unproblematic' to non-white bodies. To go unnoticed, to exist unproblematically, has an emotional dimension to it. Ahmed characterises the emotional configuration of whiteness as *comfort*: "White bodies are comfortable *as they inhabit spaces that extend their shape*. (…) Those spaces are lived as comfortable as they allow bodies to fit in" (Ahmed, 2007: 158).

Comfort, as Ahmed argues, only becomes noticeable once we lose it. This may help to explain why it is difficult to acknowledge the emotional aspects of privilege. Comfort tends to be felt with less awareness, and therefore, perhaps, with less intensity. We can see here an additional affective dimension of white ignorance as a meta-ignorance, namely, an ignorance that is unaware of itself.[8]

Comfort has other emotional ramifications: confidence, entitlement, and overconfidence or arrogance. Feelings of entitlement carry with them an unquestioned assumption about the self as capable and rightfully deserving 'by default' of whatever it desires and undertakes. They may be amongst the emotional phenomena that go most undetected by those who experience them, precisely because they presuppose these unquestioned assumptions. Ahmed's phenomenological description of whiteness as marked by *comfort*, which is fertile ground for feelings of *entitlement to comfort*, sheds light on the cluster of emotional phenomena that underpin the motivations of whites as a group.

Comfort and feelings of entitlement as the emotional structures of white privilege become even more evident when comfort is withdrawn. When white privilege is challenged, discomfort rises to the surface, and strong emotional defensive mechanisms get triggered to maintain power. This is visible in documented white defensive emotional reactions to *racial discomfort*. As a feature of the emotional structure of privilege, comfort is manifested in the display of defensive emotional reactions that seek to restore it as a way of recentering power (Applebaum, 2010).[9]

Projective Mechanisms

There is an additional relational dimension to white comfort. If white bodies are marked by feelings of comfort and entitlement, on the flip side, non-white bodies are read as that which cause discomfort. More radically, Yancy argues that the relational aspect of whiteness means that whites' sense of comfort, security, and self-esteem, is largely *dependent* on the construction of non-white bodies as defective and dangerous. Yancy contends that whites have *needed* to build their domination on the material degradation and imaginative misconstrual of the non-white body. This process involves forms of emotional parasitism. These are visible in two main mechanisms: forms of psychological projection and demands for "emotional labour" (Hochschild, 2012).

As a way of showing the relational dimension of the meaning of Blackness, Yancy analyses a common, "peculiar experience", which he names "the elevator effect":

> Well-dressed, I enter an elevator where a white woman waits to reach her floor. She 'sees' my Black body, though not the same one I have seen reflected back to me from the mirror on any number of occasions. Buying into the myth that one's dress says something about the person, one might think that the markers of my dress (suit and tie) should erase her tension. What is it that makes the markers of my dress inoperative? She sees a Black male body 'supersaturated with meaning, as they [Black bodies] have been relentlessly subjected to [negative] characterization by newspapers, newscasters, popular film, television programming, public officials, policy pundits and other agents of representation'. Her body language signifies, 'Look, *the* Black!' On this score, through a sort of performative locution, her body language functions as an insult. Over and above how my body is clothed, she 'sees' a criminal, she sees me as a threat.
>
> (Yancy, 2017: 20–21)

From his description, we can have a sense of how whites sustain oppression by displaying anxiety and fear at the mere sight of a Black person. In many instances, the Black body seems to be experienced by the white body as that which *causes* discomfort in the forms of fear and anxiety. These emotional responses to the presence of non-white bodies are, in Yancy's view, forms of projection through which whites sustain their sense of self. The image of the non-white as embodying criminality, dirtiness, disease, etc., insofar as it is placed outside the white ego, sustains the imaginary identification of the self as innocent, clean, pure, etc.

Yancy highlights the parasitic dimension of these common white emotional reactions: "[the white woman is] unaware of how the feeling of her white bodily upsurge and expansiveness is *purchased at the expense* of my Black body" (Yancy, 2017: 21. My italics). This is so because, even though this encounter with the white woman's fear "does not shatter [his identity]", it nevertheless produces an acute awareness of his own embodied existence: "I now begin to calculate, paying almost neurotic attention to my body movements, making sure that this 'Black object', what now feels like an appendage, a weight, is not too close, not too tall, not too threatening" (Yancy, 2017: 33). The emotions displayed in cases like the white woman in the elevator have an exploitative character. Comfort needs to be sustained in the face of fear. Therefore, the oppressed are implicitly required to perform different kinds of emotional labour, such as adapting their behaviour in order to reduce white discomfort: to bow in order to appear less threatening, to smile, but not enough to seem menacing, and so on.[10]

Conclusion

In this chapter, I have argued that the recalcitrant aspect of white ignorance cannot be adequately explained by only focusing on its cognitive dimensions. I have argued that what Mills calls "motivational group interest" stands apart from the other four cognitive processes that generate and sustain white ignorance. The *conative* dimension of motivational group interest suggests that it plays a major role in the recalcitrance of white ignorance. I have aimed to expand Mills' notion of motivational group interest by exploring its affective (embodied and emotional) roots, drawing on the works of Medina, Lugones, Sullivan, Ahmed, and Yancy. If such affective dimensions of white ignorance partly account for its recalcitrance, we need more than reforms at the cognitive level: affects, emotions need to be engaged in order to undo white ignorance. In this sense, if white ignorance is an important source where the coloniality of many political concepts is produced and reproduced, then any decolonial project will also need to contend with the conative and affective patterns that contribute to the formation and persistence of these concepts.

Notes

1 This is not to say, however, that whites manifest it in a uniform way:

> Whites are not a monolith, and if the analysis of white ignorance is to be part of a social epistemology, then the obvious needs to be remembered -that people have other identities beside racial ones, so that whites will be divisible by class, gender, nationality, religion, and so forth, and these factors will modify, by differential socialization and experience, the bodies of belief and the cognitive patterns of the sub-populations concerned.
>
> (Mills, 2007: 22–23)

2 The day commemorates the arrival of Christopher Columbus to the Americas in 1492.

3 https://www.esmadrid.com/agenda/desfile-12-octubre-plaza-cuzco-nuevos-ministerios retrieved on 16/01/2022.
4 Sullivan prefers to think of habits of white privilege as *unconscious*, rather than simply nonconscious or preconscious because the latter notions do not fully account for habits' "strong resistance to conscious recognition" (Sullivan, 2006: 6).
5 The distinction between a white supremacist regime where racist beliefs, behaviour, laws, etc., are consciously held does not mean, of course, that there are no unconscious habits or mechanisms particular to that era. Sullivan's argument consists in thinking the passage from white supremacist (*de jure* racist regime) to white privileged (de facto racism) as involving particular forms of repression through which many of those explicit, overt patterns have become unconscious.
6 Because habits are both constitutive of and constituted by an environment, they will also be different in each society, depending on a series of other factors which also shape the social context (institutions, religion, language, and other cultural features). Analyses such as Sullivan's or Yancy's, mainly concerned with American (United States) society, cannot simply be applied to other societies and cultures without taking into consideration these differences in context.
7 As Sullivan notes, the use of the term "seduction" does not characterise an abusive sexual act from a parent to a child:

> I immediately must clarify the term 'seduction' since it does not mean that a sexually abusive act takes place between adult and infant [...] [T]he event of seduction involves the transference of enigmatic messages about sexuality from adult to child, not a sexual act in the customary sense of the term.
>
> (Sullivan, 2006: 64)

8 By contrast, the experience of being oppressed may appear as more evidently *emotional* because the cluster of negative, painful, and disempowering emotions are probably more intensely experienced in general than what we may describe as the emotional ramifications of comfort: confidence, entitlement, and overconfidence or arrogance.
9 For example, in pedagogical contexts, people of colour are often put under pressure "to mollify white discomfort at the sacrifice of their own educational and emotional needs" (Applebaum, 2017: 867).
10 In response to the objection that the white woman's intentions may have been misinterpreted, Yancy notes that, even if we hypothetically concede the possibility of misinterpretation, the elevator example condenses a multitude of signals that have become part of a shared knowledge among Black people. Even granting the possibility of making an error of judgement in reading the white woman's expressions of fear and disgust, this does not disprove "the warranted assertability of other claims regarding the racist actions of whites" (Yancy, 2017: 24) profusely documented by non-whites as part of their common experience or racism.

References

Ahmed, S. (2007). A Phenomenology of Whiteness. *Feminist Theory*, *8*(2), 149–168. https://doi.org/10.1177/1464700107078139

Applebaum, B. (2010). *Being White, Being Good.* Lanham: Lexington Books.

Applebaum, B. (2017). Comforting Discomfort as Complicity: White Fragility and the Pursuit of Invulnerability. *Hypatia*, *32*(4), 862–875.

Bailey, A. (2007). Strategic Ignorance. In S. Sullivan & N. Tuana (Eds.), *Race and Epistemologies of Ignorance* (pp. 77–94). Albany: State University of New York Press.

Dotson, K. (2012). A Cautionary Tale. On Limiting Epistemic Oppression. *Frontiers*, *33*(1), 24–47.

Douglass, F. (2009). *Narrative of the Life of Frederick Douglass*. Cambridge, MA: Harvard University Press.
DuRocher, K. (2011). *Raising Racists: The Socialization of White Children in the Jim Crow South*. Lexington: The University Press of Kentucky.
Fanon, F. (2008). *Black Skin, White Masks*. London: Pluto Press.
Garcia, J. (1996). The Heart of Racism. *Journal of Social Philosophy*, *27*(1), 5–45.
Hochschild, A. (2012). *The Managed Heart* (3rd ed.). Berkeley and Los Angeles: University of California Press.
Lorde, A. (2017). *Your Silence Will Not Protect You*. London: Silver Press.
Lugones, M. (2003). *Pilgrimages/Peregrinajes. Theorizing Coalition Against Multiple Oppressions*. Lanham: Rowman & Littlefield Publishers, Inc.
Martín Alcoff, L. (2007). Epistemologies of Ignorance: Three Types. In S. Sullivan & N. Tuana (Eds.), *Race and Epistemologies of Ignorance*. Albany: State University of New York Press, 39–58.
Medina, J. (2013). *The Epistemology of Resistance*. New York: Oxford University Press.
Mills, C. (2007). White Ignorance. In S. Sullivan & N. Tuana (Eds.), *Race and Epistemologies of Ignorance*. Albany: State University of New York Press, 11–38.
Morrison, T. (1992). *Playing in the Dark. Whiteness and the Literary Imagination*. London: Picador.
Morrison, T. (2016). *The Bluest Eye* (2nd ed.). London: Vintage.
Spelman, E. (2007). Managing Ignorance. In S. Sullivan & N. Tuana (Eds.), *Race and Epistemologies of Ignorance*. Albany: State University of New York Press, 119–131.
Sullivan, S. (2006). *Revealing Whiteness. The Unconscious Habits of Racial Privilege*. Indianapolis: Indiana University Press.
Sullivan, S. (2015). *The Physiology of Sexist and Racist Oppression*. Oxford and New York: Oxford University Press.
Yancy, G. (2017). *Black Bodies, White Gazes* (2nd ed.). Lanham: Rowman & Littlefield.

3 The Idealised Subject of Freedom and the Refugee

Shahin Nasiri

Introduction

As with terms such as 'human rights', 'democracy', and 'equality', the notion of 'freedom' has an emblematic character with highly normative overtones. The Universal Declaration of Human Rights (hereinafter referred to as "the Universal Declaration") considers freedom both as a universal right and one of its founding principles (UN General Assembly, 1948). After the official abolition of slavery and the recognition of equal rights for women, the widely held assumption is that freedom is – at least formally – a universal entitlement belonging to every human being. The formal architecture of freedom seems consistent and inclusive, without logical tensions, anomalies, or otherness. Yet, once we turn our analytical focus to the perspectives of refugees, we realise immediately that this universalist thesis is not easily defensible. In this chapter, I argue that the universalist discourse of freedom is confined within the boundaries of the nation-state paradigm. This paradigm represents an exclusionary model for organising the world in the post-colonial era. As will be seen, refugeehood poses a serious challenge to the logic of the current nation-state model, which is premised on the colonial nexus of national citizenship, statehood, and territorial sovereignty.

This chapter begins by offering a critical overview of prevalent humanitarian approaches to the interrelation of refugeehood and freedom. The humanitarian approach represents a theoretical stance that, implicitly or explicitly, is built on the legal framework offered by the Refugee Convention (1951) and relating human rights instruments (hereinafter referred to as "the Conventional model"). The humanitarians conceptualise refugeehood in terms of a transitory, temporary, and exceptional status that should, ultimately, transform into citizenship, either by inclusion (naturalisation) or exclusion (repatriation). Correspondingly, they envision refugees as "victims of persecution" who should be granted protection on a temporary basis and under extraordinary circumstances. This humanitarian vantage point tends to reduce refugees to depoliticised victims whose biological life should be rescued by host countries and international aid organisations. It overlooks refugees' political subjectivity and poses several limitations to understanding refugees' practices and perceptions of freedom.

DOI: 10.4324/9781003293460-5

This chapter offers an alternative approach to rethinking the interrelation of freedom and refugeehood. By employing the notion of (non)-subjectivity, I elucidate that the meaning and significance of (un)freedom could not be reduced to static political categories or citizen-oriented conceptions. The central idea of this chapter is that refugeehood is a multi-faceted juridico-political condition that transgresses the hierarchical binaries of inclusion versus exclusion. For refugees, the meaning of freedom and unfreedom are intertwined and should be approached as a dynamic relationship characterising the state of (non)-subjectivity. As political (non)-subjects, refugees resist the exclusionary structures of unfreedom and enact their freedom and subjectivity in different phases of refugeehood.

Freedom and National Citizenship

The way the notion of 'freedom' has been framed within post-war human rights documents represents the prevalent tendency towards the juridico-political meaning of freedom in contemporary political discourse. In this regard, the Universal Declaration (1948) could be considered an exemplary model that provides clear insights into the formal enunciation of freedom. Let us, therefore, briefly examine how freedom has been articulated in this document. The term 'freedom' is almost excessively used in the Universal Declaration and appears more than 20 times in the document (UN General Assembly, 1948). This observation attests the importance of this political signifier in the post-war context.

The Universal Declaration makes two universality claims concerning the validity of freedom. First, freedom is asserted as a core political value that should be pursued and promoted in *every* nation globally (Morsink, 2009: 17–18). The framers of the Universal Declaration state that the "peoples of the United Nations have in the Charter reaffirmed their faith in fundamental human rights [...] and have determined to promote social progress and better standards of life in larger freedom" (UN General Assembly, 1948: preamble; see also Brown, 2016: 19). The Universal Declaration also reiterates that all "[...] member States have pledged themselves to achieve, in cooperation with the United Nations, the promotion of universal respect for and observance of human rights and fundamental freedoms" (UN General Assembly, 1948: preamble). Accordingly, freedom is considered a foundational political value, which serves as the precondition for the realisation of the rights that are formally declared (Morsink, 2009: 18). The Universal Declaration claims that common understanding of freedom is "of the greatest importance" and forms a prerequisite for the recognition and full realisation of "inalienable" human rights (Brown, 2016: 19; UN General Assembly, 1948: preamble).

The second universality claim of the Declaration relates to the *designation* of freedom in the sense of a universal and inalienable right (Brown, 2016: 19; Morsink, 2009: 17). According to the Universal Declaration, "all human beings are born free" and entitled to freedom, "without distinction of any kind, such as race, colour, sex, language, religion, or other political or other opinion, national or social origin, property, birth or other status" (UN General Assembly, 1948: Arts. 1 & 2). The universal quantifier in the phrase "all human beings" implies that freedom is an entitlement belonging to every human individual. As such, the categorical domain of freedom contains 'all members of the human family' and does not exclude anyone who is recognised as human.

Within the human rights discourse, these universality claims give shape to the formal structure of freedom. However, this universalist structure is built upon a conflictual juridico-political foundation. Despite the apparent universalisation of the idea of freedom, the juridico-political meaning of this political signifier is confined by the boundaries of national citizenship, statehood, and territorial sovereignty. Historically, the appearance of the notion of freedom, in the sense of a central normative value, goes hand in hand with the rise of the nation-state in the post-colonial world. As Balibar, Wallerstein, Anderson, and others maintain, the nation-state model is the prevailing mode of political organisation of society in the last few centuries (Anderson, 2016: 40–43; Balibar & Wallerstein, 1991: 87; Isin, 2002: 232–33). On the one hand, the gradual emergence of this model was the by-product of the internal collapse of colonial empires and the rise of new nation-states. On the other hand, the disintegration of colonial empires (caused by decolonisation and independence movements) gave rise to the globalisation of the nation-state paradigm. Both historical processes resulted in the idea that society should be organised by a sovereign state that represents and unifies a nation within a clearly bounded territory and highly secured borders (Agamben, 1998: 76, 2000; Balibar, 2016: 33–34; Balibar & Wallerstein, 1991: 91).

According to Douzinas and Balibar, the glorification of the modern concept of national citizenship gives rise to the double process of inclusion and exclusion. In the first place, it amounts to the inclusion of citizens in the domain of membership and their recognition as the main "beneficiary of rights" and freedoms (Balibar 2016: 16; Douzinas, 2019: 95). Yet, citizenship is not an unbounded juridico-political status to which every human being is entitled. The actual meaning and limits of this status are contextually determined and depend on social and historical conditions. As Losurdo observes, in the modern era, while free states were celebrating and enjoying their freedom and democratic citizenship, practices such as indigenous genocide, forceful assimilation, slavery, patriarchal rule, ethnic cleansing, and violent colonisations reached their apex (2014: 323–44; see also Douzinas, 2019: 64). Moreover, the equation of political membership with citizenship implies the exclusion of all non-citizens from the domain of humanity. This process turns freedom – instead of a universal right – into an exclusive status and binary marker that classifies the social world into two opposite poles: citizens versus non-citizens. This hierarchical division has, for centuries, prevented (fugitive) slaves, women, subjects of colonised territories, and present-day refugees from the domain of formal freedom and political membership (Bourke, 2013: 240; Douzinas, 2019: 95–96; Hesse, 2014; Isin, 2002: 3–4).

The Anomaly of the Paradigm

Despite the predominance of the nation-state paradigm, there are an increasing number of people around the globe who do not fit into this model. Statistical data and forecasts concerning the number of refugees and displaced people suggest that the global refugee population has doubled since 2011 (UNHCR, 2022b). In this sense, statelessness and refugeehood could be understood as an emerging crisis that probes and dismantles the idealised image of the human reflected in the concept of national citizenship (Balibar, 2016: 79–81). But what happens once a human being becomes a refugee? As Arendt and others make clear, the construction of modern nation-states concurs, historically, with the neo-colonial projects of nation-building and humanitarian conquests, large-scale processes of denationalisation and forced displacement. These processes made millions of people in the world refugee and stateless, which are epitomised by outlawing Jews, Roma and Sinti, Armenians, Palestinians, Indians, the Rohingya, and many other European and non-European nationals over the last few decades (Arendt, [1951]1973: 278–81; Douzinas, 2019: 97–8; Gündoğdu, 2015: 107).

Confronted with the subsequent practices of genocide and massacre, the Refugee Convention (1951) was designed as a legal instrument to prevent these atrocious phenomena (UN General Assembly, 1951). The Refugee Convention defines refugees as individuals who have a "well-founded fear of persecution" because of five specified grounds, i.e., race, religion, nationality, membership of a particular social group and political opinion (UN General Assembly, 1951: Art. 1). There are several international instruments relating to refugees (including the 1967 Protocol) that recognise their status and ascribe a minimum set of rights to them (UN General Assembly, 1967).[1] However, all these international instruments share a common underlying assumption: being refugee is a temporary and exceptional status, which should, at some point, transform into the status of citizenship. This viewpoint has, explicitly, been formulated by Sadako Ogata[2] in the *travaux préparatoires* concerning the intentions and scope of the Refugee Convention. As Ogata puts it,

> until an appropriate durable solution is found for them, and refugees cease to be refugees either through voluntary repatriation or legal integration (naturalization) in their new home country, it is necessary for them to be treated in accordance with internationally recognized basic minimum standards.
>
> (UNHCR, 1990: 4)

To preserve the formal unity of nationality and political membership, refugeehood has been viewed as an 'abnormal' and transitory juridico-political condition that should not "be regarded permanent" (UNHCR, 1990: 246). In doing so, the legal processes of *repatriation* (to the country of origin) and *naturalisation* (in the host country) are widely recognised and put into force (Ibid).

Still, the exponential growth of refugee population worldwide, radically, challenges this double mechanism. As of 2022, over 100 million people have fled their habitual place of residence due to failure of nation-states, institutional violence, humanitarian interventions, and international wars on terror. A significant number of these refugees spend their entire life in extended exile or protracted refugee situations. Millions of children are born into permanent statelessness and many of them spend their entire childhoods in refugee camps (Hyndman & Giles, 2018; Parekh, 2020: 5; UNHCR, 2022a: 20).

Humanitarian Approach

The global emergence of refugeehood throws the nation-state paradigm into an all-encompassing crisis on a practical and theoretical level. This crisis is being echoed in inflationary usage of hyperbolic metaphors in Western media, and everyday political discourse, such as 'refugee crisis', 'humanitarian catastrophe', 'mass movements of migrants', 'human floods from the Global South' threatening national borders, and so on (see, for instance, European Commission, 2017; Georgiou & Zaborowski, 2017). On their part, political theorists attempt to analyse and articulate a response to this emerging political phenomenon. In mainstream political discourse (especially among liberals and neo-republicans), the humanitarian approach has played an influential role in the way refugeehood is being depicted and understood.

The humanitarian approach represents a theoretical stance that, implicitly or explicitly, builds on the conceptual model offered by the Refugee Convention and relating human rights instruments. In this framework, refugees are often portrayed as vulnerable "victims of persecution" who are "temporarily" in need of "humanitarian" and "charitable" assistance from "host" countries and international aid organisations (Miller & Straehle, 2021: 21; UN General Assembly, 1951: Arts. 31 & 33). The term 'persecution' is directly associated with life-threatening unfreedom in the sense of extreme forms of coercion or domination (Hathaway, 2021a; Hathaway & Foster, 2014: 179). On this interpretation, persecution is almost synonymous with a condition that liberal and neo-republican theorists call unfreedom.[3]

The humanitarian approach is less concerned with refugees' political subjectivity and conceptualises them as victims of persecution who should be given protection on a temporary basis. The main point of disagreement among humanitarian political theorists revolves around the dichotomy of inclusion versus exclusion. The central question is whether and to what extent refugees are to be considered as formal right-holders (subject of freedom) and whether and to what extent receiving states have duties towards these (would-be) subjects (Blake, 2013; Carens, 2015: 278; Kukathas, 2017; Miller, 2016: 51–57).

First, to establish whether the human refugee qualifies for 'inclusion in the domain of freedom' (admission), refugeehood is, conceptually, distinguished from citizenship as well as from other types of migration (Carens, 2015: 192–224; Miller, 2017: 767). According to the Refugee Convention, refugeehood is an exceptional form of human displacement, which derives its legitimacy from a "well-founded fear of persecution". This fear of persecution should be substantiated by "objective" and "subjective" grounds (Hathaway & Hicks, 2005: 510–514). The Refugee Convention does not define how states are to determine whether individual applicants meet the refugee definition. As such, the receiving state is the only institution that is authorised to determine and recognise whether refugees' fear of persecution is valid (Hyndman & Giles, 2018: 2; UNHCR, 2019: 19, 2021). Those persons whose "fear of persecution" is recognised as "well-founded" could be granted a set of minimum rights or freedoms. It should, however, be noted that these rights and freedoms are categorically different from citizenship rights. As Miller contends, the scope of refugees' rights and freedoms depends on the hospitality of the receiving state whose priority is to protect the interest of its own nationals (Miller, 2016: 160, 2017). As a general rule, refugees are prevented from participating in the realm of politics and related processes, such as executive, legislative, or judicial bodies. Other civil rights which apply to refugees could "be withheld on grounds of their lack of nationality during national emergencies" (Hathaway, 2021a: 174).

To gain access to the political domain, the refugee needs to go through the legal process of 'naturalisation', which is the act of investing an alien with the status of a national in the receiving state (Carens, 2015: 47–48). Put another way, the refugee should acquire citizenship to take part in the political body as a fully recognised subject of freedom. The process of naturalisation is dependent on the degree of generosity of the receiving state and is codified by its special legislative directives (UNHCR, 1990: 246–48). In contrast to the universalist premise of the Universal Declaration, the (political) freedom of the (former) refugee (who becomes a 'naturalised citizen') is not immediate, inalienable, or manifest. Rather, this freedom is mediated by the generous intervention of the receiving state and derives from *natural* entitlements of freeborn citizens, i.e., those who possess freedom by birth (see also Carens, 2015: 21).

Second, although the state has exclusive authority to process refugee claims and establish asylum procedures at its own discretion, it has no direct or absolute obligation to open its borders and grant refugee status to all (potential) asylum seekers in question (Miller, 2013).[4] Asylum-seekers whose refugee claim (i.e., their well-founded fear of persecution) is not recognised, have no legal *ground* for residence and could, legitimately, be expelled and repatriated (Miller, 2016: 91–93). Repatriation is an internationally accepted mechanism to return rejected asylum-seekers to their country of origin (Hathaway, 2005; see also Carens, 2015: 208). As the etymological root of this term suggests, repatriation is the act of relocating the rejected and unrecognised fugitives to their own patria, i.e., the political territory where their (male) ancestors are born. In this sense, repatriation is another expression of the myth of territorial origin (nativity). A foreigner whose "fear of persecution" is not recognised has, by definition, no ground to stand upon; she has no legitimate legal *status*. The only option available for the politically *groundless* person is to be relocated to one's own natural territory. In this interpretation, repatriation is the legal counterpart of naturalisation and characterises the mythical nexus of nation, state, and territory (Agamben, 1998, 2000; Agier, 2017: 34).

Third and subsequently, the legal relationship between asylum-seekers and host countries is conditioned by the principle of territorial sovereignty. Following this principle, states are bound to respect human rights only when subjects physically come under their sovereignty and jurisdiction. As long as the refugee has not entered the territory of a particular nation-state, she has no legal basis to submit a valid refugee claim. Nonetheless, nation-states (potential duty-bearers) have no binding obligation to open their borders and provide access to their territory, in order to facilitate or admit refugee claims (Agier, 2017: 152; Spijkerboer, 2017). The principle of national sovereignty legitimates the exclusive right to exercise discretionary control over admissions (Blake, 2013; Miller, 2016: 160).[5] This principle has wide-ranging political consequences, especially, in wealthy Western nation-states. In the present-day political context, the state's exercise of sovereignty has led to very sophisticated and rigid border regimes, extravagant obsession with border management, and exorbitant investments in extraterritorial border control. These bordering practices unmask the concealed image of the ideal human (the citizen) that underlies the universalist discourse of human rights (Agier, 2017: 5–12; Mezzadra & Neilson, 2013: 279; Spijkerboer, 2017).

The main discussion between political theorists – whose theories implicitly or explicitly are based on humanitarian conceptions of refugeehood – centres on the state's discretionary power to exclude, its legitimacy, and moral limits. At one end of the spectrum, cosmopolitan scholars have argued for open borders and criticised the state's discretionary power and exclusionary practices of border control and corresponding policies (see, for instance, Abizadeh, 2012; Carens, 2015: 278; Kukathas, 2017). For example, as Carens argues, socio-economic advantages are distributed unequally and unfairly throughout history, dividing the world into rich and poor nation-states. Closed borders cause morally unfounded inequality (Carens, 2015: 226). On this view, citizenship of rich states is a privilege and resembles aristocratic prerogatives in the Middle Ages. For Carens, global inequality brings about the moral obligation, especially for rich nation-states, to promote freedom of movement and implement inclusive and compassionate border policies (2015: 227; see also Chamberlain, 2021).

At the other end of the spectrum, nationalist scholars have challenged this position and argued for unilateral border control. The main argument, here, relates to the protection of territorial sovereignty, national interests, and social cohesion (see, for instance, Blake, 2013; Miller, 2016, 2017). In the hierarchy of state's duties and obligations, the state has an immediate obligation ("perfect duty" to use neo-Kantian terminology) towards their own nationals and should prioritise their rights and interests (Miller, 2017). To fulfil this duty and given the fact that the high number of newcomers might lead to social disruption and lack of financial and public resources, the receiving state is morally legitimated to take restrictive measures to prevent refugees from entering its territory (Blake, 2013; Ekins, 2021; Miller, 2016: 74). As Miller argues, migration control should not be conceived as a *coercive* measure, as it is meant to "*prevent* those denied entry from carrying out their life-projects" (Miller, 2017: 768). Migration control is a *preventive* measure to filter migration flows and to distinguish irregular economic migrants from those who have a well-founded ground for leaving their country of residence (Ekins, 2021). On this view, the rationale behind strong border control is to protect national interests, to safeguard state sovereignty, and to alleviate "a perception of cultural threat and a sense that their home is under invasion on the part of members of the receiving society" (Miller, 2016: 160).

Despite diverging moral standards, both nationalist and cosmopolitan perspectives are framed within the double mechanism of naturalisation and repatriation. Both viewpoints refer to a single assumption. On the one hand, the human refugee (the alien) should be included and rendered *natural*. Through the hospitable mediation of the state's discretionary power, one stops being a fugitive and will, possibly, be granted citizenship (see Carens, 2015: 45–55). The naturalised fugitive is regarded *as if* she were a freeborn member of the society. On the other hand, the human refugee is territorially excluded from national frontiers or legitimately repatriated to her country of origin (Ekins, 2021; Miller, 2016, 2017). Border control and repatriation aim to convert refugees into national citizens of their own patria. In doing so, exclusion and bordering practices amount to a reverse form of naturalisation.

The Arendtian Critique

Within the humanitarian framework, it is not the *human* of human rights that substantiates our freedom. Freedom is only relevant for those who are already defined and recognised as full political subjects. It is one's bond with native territory and the persona of national citizenship that allows us to participate in the realm of freedom. In this way, the prevalent human rights regime reinforces the gap between man and citizen and often leaves refugees without guarantees (Douzinas, 2019: 64). National citizenship is a modern persona that is put on the face of the 'universal free human being'. This persona is arbitrarily and forcefully equated with political subjectivity and membership (Balibar, 2016: 73). Once cut off from the political community, the condition of the human being transforms into an abnormality, representing victimhood, redundancy, and powerlessness (Bauman, 2004; Mezzadra, 2020, Squire et al., 2021).

Refugeehood represents a normal by-product of the postcolonial world order and embodies the counter-history of nation-states whose internal borders, walls, and external fortifications are, growingly, being militarised and reinforced (Khosravi, 2010: 2–4; De Genova, 2017: 24; Mezzadra & Neilson, 2013: 145; Squire et al., 2021). Refugeehood is a juridico-political condition that is radically different from citizenship. The human condition of the refugee demonstrates the real and irreducible non-identity of citizenship and the fact of being born human. It embodies an anomalous juridico-political condition situated between the double mechanisms of exclusion and inclusion.

Directly after the adoption of the Universal Declaration, Arendt was one of the first theorists who observed and critically examined this anomaly. According to Arendt, it is too simplistic and misguided to conceive refugeehood as a transitory or temporary state applying to exceptional circumstances. On the contrary, the mass flight of refugees should be regarded as a normalised consequence of exclusionary structures of the nation-state model. As she goes on to argue, abstract human rights do not provide any ground for humans to be free in the political sense. In fact, the protection, recognition, and expression of freedom (and rights) only make sense within the boundaries of the political community (Arendt, [1951]1973: 176; see also Benhabib, 2018: 103–105; Gündoğdu, 2015: 37).

According to Arendt, freedom is socially instituted and is premised on the notion of political membership. As long as a human being is not recognised as a (full) member of a political community, it would be misguided to, meaningfully, speak of her right to freedom. Being part of the body politic is the precondition for human freedom and participation in the public domain (Arendt, 1960: 28). This precondition is famously formulated as the "right to have rights", which denotes the right of every individual to belong "to some kind of organised community" (Arendt, [1951]1973: 297–98). Arendt's critical diagnosis highlights several institutional boundaries and exclusionary structures that prevent refugees from being a formal subject of freedom.

First, in modern nation-states, the relationship between rights and duties is a symmetrical one. For something to be a (citizens') right, one needs to have a duty-bearer (i.e., the state) that could serve as the protector and guarantor of such rights. This state–citizen relationship defines the domain of rights and freedoms. By the same token, for something to be a *human* right, one needs to have a corresponding institution ('a world government') that could protect and guarantee this universal right (Arendt, [1951]1973: 298; see also Hamacher, 2014; Menke, 2007). As Arendt puts it, even "the best-intentioned humanitarian attempts to obtain new declarations of human rights from international organizations" should recognise that the existence of this global institution would contradict the defining elements of the current world order, which is organised based on the principles of state sovereignty and national citizenship (Arendt, [1951]1973: 298; see also Hamacher, 2014; Hayden & Saunders, 2019).

Second, to be able to assert one's freedom in a politically meaningful way, one needs "to live in a framework where one is judged by one's actions and opinions" in the first place (Arendt, [1951]1973: 296). In other words, one needs to be a participant in the public domain, which is the political manifestation of "the right to have rights". However, refugees and stateless people (*heimatlosen*) lack this foundational right. They have, *de jure* or *de facto*, lost their political membership (Arendt, [1951]1973; Balibar, 2016; Smith & Zhang, 2019: 132). According to Arendt, this loss exposes refugees to a state of rightlessness and unfreedom. Without guaranteeing the "right to have rights", it would, therefore, be non-sensical or even contradictory to consider refugees as subjects of freedom in the political sense (Arendt, [1951]1973: 296; Gündoğdu, 2015: 28).

Third, the Arendtian analysis sheds light on one of the key characteristics of refugees' experience of unfreedom (placelessness). In the existential sense, the refugee is someone who lacks a place in the human world. Refugeehood and statelessness go hand in hand with "loss of home", "loss of government protection" and loss of the entire social texture that embodies our "distinct place in the world" (Arendt, [1951]1973: 296). The refugee has lost her access to the public life that makes her opinions significant, and her action and speech effectively heard (Arendt, [1951]1973: 297–98; Benhabib & Nathwani, 2021: 123; Gündoğdu, 2015: 22). In extreme cases, "loss of polity" could expose refugees to situations (exemplified in "concentration-camp life") in which the very "possibility of fighting for freedom" is non-existent (Arendt, [1951]1973: 297–98).

Notably, Arendt's critique expresses the tension between the formal universalisation of freedom and the exclusionary structures of the nation-state model. Moreover, she offers important insights into institutional practices, legal constraints, and existential conditions that characterise refugees' unfreedom. Yet, despite her illuminating critique of formal human rights, Arendt gives a binary twist to the question of freedom. As Rancière observes, the Arendtian position is grounded on a sharp distinction between the domain of freedom and the domain of unfreedom and rightlessness. By the same token, refugees' lack of political community throws them into an existential domain that is free (devoid) of freedom (Rancière, 2010: 67; Schaap, 2011). By overemphasising structural limitations of the nation-state paradigm and internal paradoxes of abstract human rights, this criticism remains state-oriented and citizen-centric. It primarily limits its analytical focus to institutional mechanisms that amount to refugees' unfreedom. For the same reason, the Arendtian approach pays little attention to multifaceted dimensions of refugeehood and heterogenous practices of freedom, collective struggles, and lived experiences by which refugees exercise their agency and claim a place in the world.

Freedom and (Non)-subjectivity

As several critical scholars observe, state-oriented and citizen-centric reconstructions[6] of freedom privilege a particular notion of citizenship, which is historically shaped by exclusionary boundaries of race, gender, and nationality (Hesse, 2014; Mezzadra, 2004, 2020; Rancière, 2010; Roberts, 2015). As Barnor Hesse contends, hegemonic reconstructions of universal freedom (represented by liberal and neo-republican conceptions) are rooted in a colonial-racial distinction between "white citizen freedom and non-white, non-citizen unfreedom where the latter's realization of freedom can only be derived from the being and meaning of freedom as whiteness" (2014: 229). By the same token, these theoretical reconstructions disregard the political significance of freedom for political figures who were, historically, excluded from the domain of citizenship and political membership, including fugitive slaves, the colonised, non-citizens, and present-day refugees.

But can refugees be the subject of freedom without reducing or transforming their condition to a transitory and abnormal state? As we saw, the humanitarian framework gives a negative answer to this question. For humanitarians, refugees' degrees of rights and freedoms are derivative and conditioned by their formal inclusion in the domain of national citizenship. Refugees depend on the hospitality of receiving states, which do not have an unconditional obligation to offer access to their territory. This approach leaves no space for any political intervention other than moralised humanitarian policies supporting victims of persecution with food or shelter. This humanitarian vantage point generates a new identity for refugees and reduces them to absolute victims whose biological life should be rescued (Agier, 2010: 44; Rancière, 2010: 191; Squire et al., 2021).

As Malkki and others suggest, the concept of 'refugee' has increasingly become synonymous with fear, trauma, and victimhood (Ehrkamp et al., 2019: 117; Malkki, 1995a, 1995b; Squire et al., 2021: 25–27). Instead of addressing the root causes that amount to refugeehood, such as failure of post-colonial projects of nation-building, humanitarian interventions, international wars on terror, and emergence of ecological and economic crises, the humanitarian framework locates the *refugee problem* in the atomised bodies and minds of individuals (Agier, 2017: 151; Douzinas, 2007: 58–59, 2019: 65; Malkki, 1995a: 8). The refugee is perceived as a depoliticised person without a stable status allowing her to experience the world as the subject of freedom.

According to Mezzadra, the humanitarian image of the victim defies refugees' agency and underplays "the dense fabric of subjective attempts, efforts, tensions, needs, desires, and claims" that constitute their everyday experiences and struggles (2020: 434). Therefore, the figure of the refugee can, by no means, be reduced to derivative, transitory conditions implying social death and victimhood. Nor does refugeehood personify a dehumanised being whose existence is reduced to rightlessness. In fact, the figure of the refugee is the very expression of a deep-rooted juridico-political conflict, which unmasks and unsettles the idealised persona of the *human* in human rights. It characterises the crisis of the present-day political paradigm, which is based on colonial hierarchies between national citizens versus non-citizens and refugees.

Refugeehood is a juridico-political position that transgresses the binaries of inclusion versus exclusion and rightlessness versus citizenship. The refugee is neither included nor excluded. She is both an excluded alien as well as a rightful claimant of rights and freedoms. On the one hand, refugees are subjected to oppression, institutional violence, and bordering practices. On the other, they exercise their subjectivity and freedom through courageous and purposeful acts of flight and border crossing (Mezzadra, 2004, 2018, 2020; Tazzioli, 2021). While the refugee could be subjected to exclusionary measures, she is also the ultimate subject of resistance against acts of border-making (Celikates, 2019; Mezzadra, 2018; Tazzioli et al., 2018). In this way, refugees place themselves inside and outside of the conflictual domain of politics. They are both the subject *and* the non-subject of freedom.

To articulate this ambiguous juridico-political condition, I employ the term *(non)-subjectivity*. The notion of (non)-subjectivity represents the perspective of political figures who enact, articulate, and experience freedom by transgressing the exclusionary limits of politics and citizenship. It signifies a mode of being in which universal freedom is (structurally) denied or violated, while it is being articulated, practiced, and enacted by those who are excluded, i.e., the (non)-subjects. Correspondingly, a *political (non)-subject* is defined as someone who is formally not a bearer of freedom, yet capable of experiencing and enacting freedom in her own particular way.

Historically, the conflictual articulation of freedom by (non)-subjects could be attested by practices of marronage through which fugitive slaves articulated and exercised their freedom (Hesse, 2014; Roberts, 2015). It is also manifested in the emancipatory interventions by which revolutionary women (such as Olympe de Gouges) called the formal limits of freedom into question and opened its semantics for new interpretations (Bourke, 2013: 136; Rancière, 2010: 57). For these (non)-subjects, freedom is not a pre-given right or status that one could obtain through consensual processes. On the contrary, for those who are excluded from the formal domain of political membership, freedom does not collapse into a static entitlement or right which is either granted or denied. The meaning and significance of freedom take shape in a conflictual domain which "opens up a dispute" about what this signifier means and how it should be exercised (Rancière, 2010: 68). For (non)-subjects, freedom is the product of practices, lived experiences, and struggles by which they appear on the political scene and reclaim an equal place in the human world. Freedom manifests itself in the act of flight and border crossing, resistance against structures of exclusion (e.g., slavery, racism, patriarchy, and colonialism), and a performative call for equality.

Refugees' (Non)-subjectivity

This point of departure invites us to inquire into the significance of freedom from the perspective of (non)-subjects. Let us, therefore, rearticulate the question of freedom: can the refugee be the (non)-subject of freedom without reducing her position to one of the sides of the pole, the rightless versus the citizen? To prepare a provisional answer to this question, I shall confine myself to the legal characterisation of the term 'refugee' as defined by the Refugee Convention. For the purpose of this study, I leave, therefore, the experiential dimensions of refugeehood out of account and do not explore heterogenous practices and perceptions by which refugees give meaning to freedom and unfreedom.[7]

According to the Refugee Convention, a human being becomes a refugee when she flees her homeland *because of* her well-founded fear of persecution (UN General Assembly, 1951: art. 1). The refugee escapes from a situation in which the universal right to freedom is violated. This violation stems from refugees' *otherness* and membership in a *particular* race, religion, nationality, social group, or political conviction. In her homeland (or the country of her habitual residence), the refugee is subjected to coercion or domination and is facing life-threatening circumstances that are imposed upon her by the state or state-like actors (Goodwin-Gill, 2014: 39). Following this Conventional definition, the spatio-temporal modality of refugeehood could be divided into three intertwined conditions: a) The *pre-flight* condition that characterises the state of persecution (past), b) the condition of *flight*, which represents the physical movement of the refugee from her homeland (present), and c) the *post-flight* condition which represents the indeterminate position of the refugee in the receiving state (future). These entwined conditions constitute the meaning of refugeehood and give shape to the dynamic relationship between refugees' freedom and unfreedom.

Pre-flight condition (past): The pre-flight condition denotes the moment of persecution and the original cause for refugees' escape (Andrade, 2021: 318; Goodwin-Gill, 2014: 39). The *pre-flight* condition could be interpreted as an exemplary case for unfreedom, both in the liberal and republican sense. This condition is marked by coercion and domination and the violation of the universal right to freedom. As we have seen, this violation stems from the *otherness* of the citizen for reasons of membership in a particular social group. It bears, however, noting that the coerced (unfree) citizen is, in the strict sense, not yet to be considered a refugee. In accordance with the Conventional definition, the persecuted person becomes a refugee once she physically flees from the territory of persecution (Foster & Lambert, 2019: 37; UN General Assembly, 1951). It is the act of escape and (irregular) migratory movement, which transforms the persecuted person into a refugee. For this evident reason, the refugee could not equivocally be located in the domain of (liberal or neo-republican) unfreedom.

Flight (present): Refugeehood derives its significance from the purposeful act of flight. The condition of flight represents physical escape from the condition of coercion or domination. In and through the act of flight, the refugee *frees* herself from persecution and becomes the subject of freedom in an ambiguous manner. The act of flight transfers the human refugee to a different domain, both in the spatial and legal sense. From this point of view, modern refugeehood shows family resemblances with the condition of fugitive slaves and abolitionist practices of flight and border crossing (Mezzadra, 2020; Roberts, 2015: 170).

The condition of flight signifies the active negation of persecution, coercion, and domination. The refugee is exercising and experiencing her freedom by negating the state of persecution. Refugees' freedom is expressed in the dynamic process of flight. Flight signifies courageous practices of escape, desertion, and border crossing by which (non)-subjects negate the exclusionary determinations of borders and political structures of unfreedom. Moreover, it denotes a purposeful migratory movement in and through which (non)-subjects exercise their subjectivity and claim a place of refuge in the world (Hardt & Mezzadra, 2020; Mezzadra, 2020; Squire et al., 2021).

Post-flight condition (future): In the receiving countries, the refugee becomes a non-citizen residing in a foreign territory (Hathaway, 2021b: 171–72; UN General Assembly, 1951). Therefore, she could not be regarded as a subject of freedom in the neo-republican or Arendtian sense. Despite the physical escape, the fear of persecution remains a vital factor in the way refugees are framed in the host countries. This *fear* is the only legitimate cause for the refugee to seek asylum in the receiving state (Chetail, 2021: 205–206). In addition, the temporariness of refugee status and its causal link with fear of persecution makes her dependent on the hospitality and benevolence of the receiving states. In other words, although the refugee has freed herself from the condition of coercion, she is in need of the (benevolent) interference of the host country without which her temporary place of refuge could not be guaranteed. In fact, the non-interference of the host country leads to the abandonment and rejection of the person who is seeking refuge (see also Bauman, 2004: 76; Squire et al., 2021).

In their post-flight condition, the juridico-political instability of refugee status exposes many refugees to processes of victimisation, alienation, and institutional abandonment and unfreedom. However, refugees develop various counterstrategies and emancipatory practices by which they negate these conditions and reclaim their subjectivity and freedom. In fact, the meaning of freedom becomes intelligible and experienceable in interactions, collective struggles, and interpersonal encounters by which refugees unsettle processes of alienation, victimisation, and abandonment. As several critical scholars point out, these acts of freedom are manifested in refugees' everyday demands for equal treatment, struggles for the sake of justice, as well as collective acts of desertion and civil disobedience (Celikates, 2019; Mezzadra, 2004, 2020; Squire et al., 2021: 188–89). Evidently, under these dynamic circumstances, the meaning of freedom could not be defined and designated by way of static and state-oriented distinctions between inclusion versus exclusion, citizenship versus rightlessness.

Conclusion

As we have seen, citizen-centric and state-oriented theories of freedom disregard the dynamic nature of the question of freedom for (non)-subjects whose liberatory experiences are manifested in flight, movement, and creative resistance against acts of border-making. These theories are rooted in colonial-racial constructions of state–citizen relationship, which, historically, excluded fugitive slaves, the colonised, and non-citizens from the domain of citizenship, freedom, and political subjectivity. Correspondingly, these static interpretations marginalise the perspective of refugees (and other people on the move) whose practices and projects of freedom take shape by transgressing the boundaries of state–citizen relationships.

In different spatio-temporal phases of refugeehood, freedom and unfreedom act as two decisive factors and give shape to a dynamic relationship. Conventionally speaking, the origin of refugeehood consists in the violation of the universal (formal) right to freedom by state or state-like actors. This violation stems from the persecution (i.e., coercion or domination) of the human being in question in her habitual place of residence. However, the persecuted person should not, simply, be conflated with refugees. The persecuted person becomes a refugee by fleeing from the state of unfreedom. Flight encompasses practices of escape, border crossing, and physical movement by which refugees negate the state of unfreedom in their pre-flight condition.

Consequently, refugeehood could not (even in its Conventional sense) be explained by distinctions between exclusion versus inclusion, national citizenship versus rightlessness. Instead, refugees should be regarded as political (non)-subjects who call the formal limits of freedom into question. As political (non)-subjects, refugees interfere in the hierarchies of citizens versus non-citizens, subjects versus non-subjects. They disentangle freedom from its colonial-racial boundaries and, as a result, unlock its semantic field for new interpretations and signifying practices. Viewed from the perspective of refugees, freedom is not a totalising experience or static state, which one enjoys in the absence of coercive or dominating factors, nor is it reducible to negative entitlements or guarantees granted by the state to its citizens. On the one hand, freedom manifests itself in the active negation of the state of unfreedom (persecution or abandonment). On the other hand, freedom is enacted in the creative process of movement and border crossing by which refugees claim an equal place of refuge in the world. Therefore, refugees' freedom is, first and foremost, expressed in the act of flight and border crossing, resistance against acts of border making, struggles for justice and equality, and other emancipatory signifying practices by which they claim an equal place in the world.

Notes

1 Protocol relating to the Status of Refugees (1967).

2 Sakago Ogata was the head of the United Nations High Commissioner for Refugees (UNHCR) from 1991 to 2000.

3 For a discussion of liberal and neo-republican conceptions of freedom, see Berlin ([1958] 2008), Kramer (2008), Nasiri (2022) or Pettit (2016).

4 To be recognised as (Convention) refugees, asylum seekers undergo sophisticated migration procedures, hearings, and bureaucratic regulations that are put into force by receiving countries in view of their priorities and domestic interests.

5 The principle of national sovereignty, apart from its historical dimension, is incorporated in Article 2 of the Charter of the United Nations (drafted 1941, entered into force 1945).

6 These are 'reconstructions' as they are formulated in view of classical/ancient interpretations of freedom. Many of these theories revive certain republican/democratic discussions on freedom (e.g., neo-republican reconstructions). Or, they are developed as a critical response to those (e.g., liberal reconstructions).

7 It should, however, be noted that this Conventional definition is too narrow and too reductionistic to account for mass flight of refugees in the 21st century. Refugees' motivations for flight are shaped by various social, political, economic, and environmental factors and cannot be reduced to this framework (see also Agier, 2010; Crawley & Skleparis, 2018; Squire et al., 2021).

References

Abizadeh, A. (2012). On the Demos and Its Kin: Nationalism, Democracy, and the Boundary Problem. *American Political Science Review*, *106*(4), 867–882.

Agamben, G. (1998). *Homo Sacer: Sovereign Power and Bare Life*. Trans. By D. Heller-Roazen. Stanford: Stanford University Press.

Agamben, G. (2000). *Means without End: Notes on Politics*. Trans. By V. Binetti, & C. Cesarino. Minneapolis: University of Minnesota Press.

Agier, M. (2010). *On the Margins of the World: The Refugee Experience Today*. Cambridge: Polity.

Agier, M. (2017). *Borderlands: Towards an Anthropology of the Cosmopolitan Condition*. Cambridge: Polity.

Anderson, B. (2016). *Imagined Communities: Reflections on the Origin and Spread of Nationalism*. London: Verso.

Andrade, J. F. (2021). Regional Refugee Regimes: Latin America. In C. Costello, M. Foster, & J. McAdam (Eds.), *The Oxford Handbook of International Refugee Law* (pp. 315–333). Oxford: Oxford University Press.

Arendt, H. (1960). Freedom and Politics: A Lecture. *Chicago Review*, *14*(1), 28–46.

Arendt, H. ([1951]1973). *The Origins of Totalitarianism*. New York: Harcourt Brace Jovanovich.

Balibar, E. (2016). *Citizenship*. Trans. By T. Scott-Railton. Cambridge: Polity Press.

Balibar, E., & Wallerstein, I. M. (1991). *Race, Nation, Class*. London: Verso.

Bauman, Z. (2004). *Wasted Lives: Modernity and Its Outcasts*. Cambridge: Polity Press.

Benhabib, S. (2018). *Exile, Statelessness, and Migration: Playing Chess with History from Hannah Arendt to Isaiah Berlin*. Princeton, NJ: Princeton University Press.

Benhabib, S., & Nathwani, N. (2021). The Ethics of International Refugee Protection. In C. Costello, M. Foster, & J. McAdam (Eds.), *The Oxford Handbook of International Refugee Law* (pp. 114–133). Oxford: Oxford University Press.

Berlin, I. ([1958]2008). Two Concepts of Liberty. In H. Hardy (Ed.). *Liberty: Incorporating Four Essays on Liberty* (pp. 118–172). Oxford: Oxford University Press.

Blake, M. (2013). Immigration, Jurisdiction, and Exclusion. *Philosophy & Public Affairs*, *41*(2), 3–130.

Bourke, J. (2013). *What It Means to be Human: Reflections from 1791 to the Present*. London: Virago.

Brown, G. (2016). *The Universal Declaration of Human Rights in the 21st Century: A Living Document in a Changing World*. New York: NYU Global Institute for Advanced Study.

Carens, J. H. (2015). *The Ethics of Immigration*. New York: Oxford University Press.

Celikates, R. (2019). Constituent Power Beyond Exceptionalism: Irregular Migration, Disobedience, and (Re-)constitution. *Journal of International Political Theory*, *15*(1), 67–81.

Chamberlain, J. A. (2021). Challenging Borders: The Case for Open Borders with Joseph Carens and Jean-Luc Nancy. *Journal of International Political Theory*, *17*(3), 240–256.

Chetail, V. (2021). Moving towards an Integrated Approach of Refugee Law and Human Rights Law. In C. Costello, M. Foster, & J. McAdam (Eds.), *The Oxford Handbook of International Refugee Law* (pp. 202–221). Oxford: Oxford University Press.

Crawley, H. & Skleparis, D. (2018). Refugees, Migrants, Neither, Both: Categorical Fetishism and the Politics of Bounding in Europe's "Migration Crisis". *Journal of Ethnic and Migration Studies*, *44*(1), 48–64. https://doi.org/10.1080/1369183X.2017.1348224

De Genova, N. (2017). The Borders of "Europe" and the European Question. In N. De Genova (Ed.), *The Borders of Europe: Autonomy of Migration, Tactics of Bordering* (pp. 1–37). Durham: Duke University Press.

Douzinas, C. (2007). *Human Rights and Empire: The Political Philosophy of Cosmopolitanism*. Abingdon: Routledge.

Douzinas, C. (2019). *The Radical Philosophy of Rights*. Abingdon: Routledge.

Ekins, R. (2021). The State's Right to Exclude Asylum-Seekers and (Some) Refugees. In D. Miller & C. Straehle (Eds.), *The Political Philosophy of Refuge* (pp. 39–58). Cambridge: Cambridge University Press.

Ehrkamp, P., Loyd, J. M., & Secor, A. (2019). Embodiment and Memory in the Geopolitics of Trauma. In. K. Mitchell, R. Jones, & J. L. Fluri (Eds.), *Handbook on Critical Geographies of Migration* (pp. 117–126). Cheltenham, UK: Edward Elgar Publishing.

Foster, M., & Lambert, H. (2019). *International Refugee Law and the Protection of Stateless Persons*. Oxford: Oxford university press.

European Commission (2017). *Migration on the Central Mediterranean Route Managing Flows, Saving Lives*. High Representative for Foreign Affairs and Security Policy. Retrieved from https://eur-lex.europa.eu/legal-content/EN/TXT/?uri=JOIN%3A2017%3A4%3AFIN

Georgiou, M., & Zaborowski, R. (2017). *Media Coverage of the 'Refugee Crisis': A Cross-European Perspective*. Strasbourg: Council of Europe. Retrieved from https://edoc.coe.int/en/refugees/7367-media-coverage-of-the-refugee-crisis-a-cross-european-perspective.html

Goodwin-Gill, G. S. (2014). The International Law of Refugee Protection. In E. Fiddian-Qasmiyeh, G. Loescher, K. Long, & N. Sigona (Eds.), *The Oxford Handbook of Refugee and Forced Migration Studies* (pp. 36–47). Oxford: Oxford University Press.

Gündoğdu, A. (2015). *Rightlessness in an Age of Rights: Hannah Arendt and the Contemporary Struggles of Migrants*. New York: Oxford University Press.

Hamacher, W. (2014). On the Right to Have Rights: Human Rights; Marx and Arendt. *CR: The New Centennial Review*, *14*(2), 169–214. https://doi.org/10.14321/crnewcentrevi.14.2.0169

Hardt, M., & Mezzadra, S. (2020). Introduction: Migrant Projects of Freedom. *The South Atlantic Quarterly*, *119*(1), 168–175. https://doi.org/10.1215/00382876-8007853

Hathaway, J. C. (2005). The Right of States to Repatriate Former Refugees. *Ohio State Journal on Dispute Resolution*, *20*(1), 175–216.

Hathaway, J. C. (2021a). *The Rights of Refugees under International Law*. New York: Cambridge University Press.

Hathaway, J. C. (2021b). The Architecture of the UN Refugee Convention and Protocol. In C. Costello, M. Foster, & J. McAdam (Eds.), *The Oxford Handbook of International Refugee Law* (pp. 171–186). Oxford: Oxford University Press.

Hathaway, J. C., & Foster, M. (2014). *The Law of Refugee Status*. West Nyack: Cambridge University Press. https://doi.org/10.1017/CBO9780511998300

Hathaway, J. C., & Hicks, W. S. (2005). Is there a subjective element in the refugee convention's requirement of 'well-founded fear'? *Michigan Journal of International Law*, *26*(2), 505–562.

Hayden, P., & Saunders, N. (2019). Solidarity at the Margins: Arendt, Refugees, and the Inclusive Politics of World-Making. In K. Hiruta (Ed.), *Arendt on Freedom, Liberation, and Revolution* (pp. 171–200). Cham: Palgrave Macmillan.

Hesse, B. (2014). Escaping Liberty: Western Hegemony, Black Fugitivity. *Political Theory*, *42*(3), 288–313. https://doi.org/10.1177/0090591714526208

Hyndman, J., & Giles, W. (2018). *Refugees in Extended Exile: Living on the Edge*. New York: Routledge.
Isin, E. (2002). *Being Political: Genealogies of Citizenship*. Minneapolis: University of Minnesota Press.
Khosravi, S. (2010). *"Illegal" Traveller: An Auto-Ethnography of Borders*. London: Palgrave Macmillan UK.
Kramer, M. H. (2008). *The Quality of Freedom*. Oxford: Oxford University Press.
Kukathas, C. (2017). On David Miller on Immigration Control. *Critical Review of International Social and Political Philosophy*, *20*(6), 712–718. https://doi.org/10.1080/13698230.2016.1231833
Losurdo, D. (2014). *Liberalism: A Counter-History*. London: Verso.
Malkki, L. H. (1995a). Refugees and Exile: From 'Refugee Studies' to the National Order of Things. *Annual Review of Anthropology*, *24*, 495–523.
Malkki, L. H. (1995b). *Purity and Exile: Violence, Memory, and National Cosmology among Hutu Refugees in Tanzania*. Chicago: University of Chicago Press.
Menke, C. (2007). The 'Aporias of Human Rights' and the 'One Human Right': Regarding the Coherence of Hannah Arendt's Argument. *Social Research*, *74*(3), 739–762.
Mezzadra, S. (2004). The right to escape. *Ephemera*, *4*(3), 267–275.
Mezzadra, S. (2018). In the Wake of the Greek Spring and the Summer of Migration. *South Atlantic Quarterly*, *117*(4), 925–933.
Mezzadra, S. (2020). Abolitionist Vistas of the Human. Border Struggles, Migration, and Freedom of Movement. *Citizenship Studies*, *24*(4), 424–440. https://doi.org/10.1080/13621025.2020.1755156
Mezzadra, S., & Neilson, B. (2013). *Border as Method, or, the Multiplication of Labor*. Durham: Duke University Press.
Miller, D. (2013). Border Regimes Human Rights. *Law and Ethics of Human Rights*, *7*(1), 1–23
Miller, D. (2016). *Strangers in Our Midst: The Political Philosophy of Immigration*. Cambridge: Harvard University Press.
Miller, D. (2017). Migration and Justice: A Reply to My Critics. *Critical Review of International Social and Political Philosophy*, *20*(6), 763–773.
Miller, D., & Straehle, C. (2021). *The Political philOsophy of Refuge*. Cambridge: Cambridge University Press.
Morsink, J. (2009). *Inherent Human Rights Philosophical Roots of the Universal Declaration*. Philadelphia: University of Pennsylvania Press.
Nasiri, S. (2022). Freedom and the Imaginary Dimension of Society. *Iranian Yearbook of Phenomenology*, *1*(1), 217–238. https://doi.org/10.22034/iyp.2020.239890
Office of the United Nations High Commissioner for Refugees. (2019). *Global trends: Forced Displacement in 2018*. Switzerland: UNHCR. Retrieved from https://www.unhcr.org/statistics/unhcrstats/5d08d7ee7/unhcr-global-trends-2018.html
Parekh, S. (2020). *No refuge: Ethics and the Global Refugee Crisis*. New York: Oxford University Press.
Pettit, P. (2016). The Globalized Republican Ideal. *Global Justice*, *9*(1). https://doi.org/10.21248/gjn.9.1.101
Rancière, J. (2010). *Dissensus: On Politics and Aesthetics*. Trans. By S. Corcoran. London: Continuum.
Roberts, N. (2015). *Freedom as Marronage*. Chicago: University of Chicago Press.

Schaap, A. (2011). Enacting the Right to Have Rights: Jacques Ranciere's Critique of Hannah Arendt. *European Journal of Political Theory*, *10*(1), 22–45.

Spijkerboer, T. (2017). Wasted Lives: Borders and the Right to Life of People Crossing Them. *Nordic Journal of International Law*, *86*(1), 1–29. https://doi.org/10.1163/15718107-08601004

Squire, V., Perkowski, N., Stevens, D. & Vaughan-Williams, N. (2021). *Reclaiming Migration: Voices from Europe's 'Migrant Crisis'*. Manchester: Manchester University Press.

Smith, W., & Zhang, S. (2019). Resisting Injustice: Arendt on Civil Disobedience and the Social Contract. In K. Hiruta (Ed.), *Arendt on Freedom, Liberation, and Revolution* (pp. 115–138). Cham: Palgrave Macmillan.

Tazzioli, M. (2021). Towards a Genealogy of Migrant Struggles and Rescue. The Memory of Solidarity at the Alpine Border. *Citizenship Studies*, *25*(5), 603–619. https://doi.org/10.1080/13621025.2021.1923657

Tazzioli, M., Garelli, G., & De Genova, N. (2018). Autonomy of Asylum?: The Autonomy of Migration Undoing the Refugee Crisis Script. *South Atlantic Quarterly*, *117*(2), 239–265.

UN General Assembly. (1948). *Universal Declaration of Human Rights*, 10 December 1948, 217 A (III), available at: https://www.refworld.org/docid/3ae6b3712c.html [accessed 17 January 2022]

UN General Assembly. (1951). *Convention Relating to the Status of Refugees*, 28 July 1951, United Nations, Treaty Series, vol. 189, p. 137, available at: https://www.refworld.org/docid/3be01b964.html [accessed 12 October 2021]

UN General Assembly. (1967). *Protocol Relating to the Status of Refugees*, 31 January 1967, United Nations, Treaty Series, vol. 606, p. 267, available at: https://www.refworld.org/docid/3ae6b3ae4.html [accessed 18 October 2021]

UN High Commissioner for Refugees. (1990). *The Refugee Convention, 1951: The Travaux Préparatoires Analysed with a Commentary by Dr. Paul Weis*, available at: https://www.refworld.org/docid/53e1dd114.html [accessed 14 January 2022]

UN High Commissioner for Refugees. (2021). *Handbook on Procedures and Criteria for Determining Refugee Status and Guidelines on International Protection under the 1951 Convention and the 1967 Protocol Relating to the Status of Refugees*, April 2019, HCR/1P/4/ENG/REV. 4, available at: https://www.refworld.org/docid/5cb474b27.html [accessed 12 October 2021]

UN High Commissioner for Refugees. (2022a). *Global Trends Report 2021*. Retrieved from https://www.unhcr.org/62a9d1494/global-trends-report-2021 [Accessed 26 July 2022]

UN High Commissioner for Refugees. (2022b). *UNHCR: Global Displacement Hits Another Record, Capping Decade-Long Rising Trend.* Retrieved from https://bit.ly/3zFAgzI [Accessed 26 July 2022]

Part II

Feeling Coloniality

Bodies, Sexuality, and Agency

4 Politics without a Proper Locus

Political Agency between Action and Practice

Henrike Kohpeiß and Marie Wuth

Introduction: Smoke Onstage

In her 1964 interview with Günter Gaus, Hannah Arendt makes an appearance that is iconic to this day (1964).[1] She smokes, listens to the questions patiently, and rejects some of them as uninformed or redundant with a determined, yet charming attitude. She controls the situation with effortless intellectual authority over the man in the opposite chair. Right at the beginning of this interview, Gaus is eager to attach an unloved label to her. In his eyes, she is a philosopher – a profession, he adds, "some might regard as a masculine one". "I don't belong to the circle of philosophers", Arendt casually replies. But she also states that of course "it is possible that one day a woman will be a philosopher". When Gaus intervenes and repeats his evaluation of her person, possibly to flatter her, she accepts this ascription in a manner that signals her indifference to it: "Well, I can't help that". Arendt's reason to prefer "political theory" as a description for her job is the burden of specific interests that political philosophy has carried "since Plato" and that doesn't allow for the philosopher to "be neutral with regard to politics". "I want to see politics with an eye unclouded by philosophy", she declares. Unclouded? How can Arendt's gaze be unclouded when the smoke of cigarettes coats her sight and appears to be a condition of her thought?

Somewhere else,[2] in 2020, Harmony Holiday and Fred Moten talk about Miles Davis' onstage and offstage performances (2020). They invent the term "blackstage" to describe what happens in between the show and the continuation of it when the focus shifts. In a collective stage performance, in jazz, nothing is ever backstage, it is, at best, blackstage, which means not produced for the eyes of an audience but still perceivable and therefore part of the show or of some show. It often can appear, Harmony Holiday remarks in the case of Miles Davis, "quite macho". With regard to Billie Holiday and reference to Fumi Okiji who studies her movements, Moten says that "to hear her, in a certain sense, is also to see her" and this might well be true for Arendt. Arendt performs intellectual clarity in the midst of smoking dust. She accepts to be seen as a philosopher while she refuses to describe herself as one in the name of clarity. Clearly, Arendt is on stage while saying that, but her smoking is a

DOI: 10.4324/9781003293460-7

transition to the backstage that has not been banned from the audience's eyes. And as an audience, we see her speaking with a clouded gaze that she visibly tries to clarify with words.

Until today, Arendt is one of the most influential references for our understanding of action and politics. But her theory is also part of a 'Western' tradition of political thinking that is pervaded by the spirit of European Enlightenment and centred on the autonomous, self-determined, and independent subject. In this chapter, we will argue that the predominantly 'Western' understanding of agency strategically and structurally deprives subjects who are suffering absolute forms of domination. As Saidiya Hartman (1997) has pointed out, the position of the subjected is rather being disguised than liberated when the category of self-determined subjectivity is being applied to them. Hence, considering the absolute domination and constraint that have determined the lives of the enslaved, action looks quite different and agency appears as an almost meaningless term. The task is, thus, to examine and redefine agency with regard to the enduring coloniality of power inherent to this concept.

In this chapter, we aim to develop a decolonial perspective on agency that emphasises the situatedness and relationality of agents and allows to take their capacities as well as their experiences and hindrances into account. Agency, in this perspective, is a controversial political concept precisely because it seems to mark a vantage point of political struggle. It seems that either is agency to be rehabilitated as a key term for political subjectivity by being conceptualised in a more inclusive manner, or it is to be abandoned as an individualised concept and to be replaced by a more collective idea of political action.

The chapter aims to sketch a path for agency that values its conceptual force in Arendt's description of the political but also recognises its inaccessibility and idealistic implications about freedom that makes it appear less eligible for envisioning structural changes. Taking a closer look at the conceptual implications on each side will help to resolve the simplifying alternative that forces to choose between rehabilitation rejection. Agency can, after all, be a legitimate political claim and at the same time not serve as a given capacity of political subjects. We will contrast Arendt's concept of action with Saidiya Hartman's concept of practice by using the 'split gaze' of Fred Moten (2017, 2018). It will help us to account for the differences of intellectual paradigms leading to different interests in and assumptions about political subjectivity. Engaging with Hartman's concept of practice is part of our endeavour to open up a decolonial perspective on agency. Agency is contrasted by practice which is not bound to form or position but a capacity of relation. Hartman's (1997) concept of practice is always specific and takes into account the conditions, capacities and relations under which it emerges and is performed. Therefore, reading these theories together enables us to reflect on the particular and specific ground on which concepts and their critique grow and, overall, forges a new angle on agency.

In this chapter, we aim to challenge political agency and action by unravelling the oppressive and limiting structures inherent in the theoretical tradition and conceptualisation. Therefore, we will first turn to Arendt and discuss her ambitions and the cornerstones of her theory of political action and agents. Then, we shall focus on Moten's critical yet charitable read of Arendt's work, especially his study of her gaze and judgements. Against this background, we shall introduce Hartman's understanding of practice and critical discussion of the category of self-determined subjectivity that is vital for our decolonial approach to agency. In the final part, we will re-read agency alongside practice and turn to the aesthetic aspects of these notions. The paradigms presented in this chapter will not be fully reconciled with one another but will only meet in the idea of practice to then part ways again.

Arendt Experiences Loss

For Arendt "life without speech and action [...] is literally dead to the world" (1958: 176). Words like these express Arendt's concern about the rise of the social and demise of the political. In the 1950s, Arendt is deeply perturbed by the world loss of the 'modern man' who, according to Arendt's appraisal, regards the earth as a prison and who has forgotten what it means to be 'truly active' (1958). This loss of the world is tantamount to the loss of political life that is manifest in the inability to act, the only genuine human activity. Arendt's emphasis on action is an attempt to retrieve the political realm from the ruins of European totalitarianism left. Totalitarian and tyrannical regimes render the public space of the political, that is 'proper politics', impossible. Politics lives from the fact that people can appear as who they are and can set a new beginning through action. With word and deeds, humans take initiative and insert themselves in the web of human affairs, which is why Arendt compares action to birth (1958: 178). Thus, she introduces the concept of natality into her political theory to highlight that action is part of the human condition and that it is vital for humans to actively express their uniqueness so that they can experience the plurality of existence. The search for a pure and authentic political, thus, is more than a return to the conditions of human activity and existence; it is the attempt to acknowledge human plurality and ensure civil equality in a sphere of freedom. Freedom, as imagined by Arendt, is the goal of politics and is to be realised through action in concert with others.

Politics and Action as an Answer to Loss

Arendt's answer to that world loss is the formulation of an emphatic notion of action entwined with a protectionist and exclusive notion of the political. This answer is, however, not a signpost, it is not a vision for future politics, and, importantly, it is not demanding to become efficacious. Even though it is almost impossible not to conceive of her praise of antiquity as an ideal paradigm when she contrasts it with her harsh criticism and unapologetic appraisal of our contemporary political disasters and incapacities to be 'proper agents'.

The daring that her thinking without rails represents does not consist in a courageously activist and especially not in a somewhat seditious agenda, but in the will to understand. Arendt is writing to understand and, therefore, she is telling the story of the transformation of political life and the public sphere (Benhabib, 1992: 91). Her writings are a walkthrough of the conditions and limits of political action and behaviour, declaring the meaning of politics to be freedom.

Arendt has a specific place in mind when she writes about truly political politics, that is, the Greek *polis*. Her model of the *vita activa* mirrors the cartography of the *polis* by drawing clear demarcation lines between the private, social, and political realms and by delimiting the public and the private sphere. Thus, the political is detached from social questions and must keep up the fences enclosing the private. Neither discrimination, characteristic of the social realm, nor necessities, dominating the private life, have a place in the political arena where men come together as equals and can act in concert. Arendt's ambitious understanding of action is tied to an image of the public sphere where people can insert themselves with words and deeds in the net of human affairs (1958: 176ff). The locus of action is the net of human relations, "with its innumerable, conflicting wills and intentions" (1958: 184). Thus, human plurality and togetherness are the conditions of any action as action needs to be seen, performed, and heard (1958: 175). Only when we reveal ourselves with words and deeds in the light of the public, we can be political agents and be 'truly human':

> What makes man a political being is his faculty of action; it enables him to get together with his peers, to act in concert, and to reach out for goals and enterprises that would never enter his mind, let alone the desires of his heart, had he not been given this gift-to embark on something new.
> (Arendt, 1970: 82)

Uncovering 'the archetype' of political action in the life of the *polis* is an attempt to situate action in a model of participatory democracy that also highlights the importance of sharing opinions and the power arising when people come together as equals. Arendt has often been portrayed as a proponent of an associative model of politics (Marchart, 2007). Importantly, however, as has been pointed out by scholars like Seyla Benhabib (1992) and Bonnie Honig (1992, 1993), there is also an undeniable agonistic element. Agonism is already inscribed in the Greek model which is so crucial for Arendt's theory of action and the political, respectively. The political is a zone of competition permeated by the agonal spirit to prove oneself and reveal one's unique distinctness to others (Arendt, 1958: 194). The agonal spirit, which Arendt finds in the deeds of Greek heroes like Achilles and Hector, is a passionate drive that can be translated as the will to self-revelation and self-appearance, which accompanies appearance and action. The *agon* is a structuring moment for the political though this realm is not marked by combat or hostility.[3] After all, what characterises the political is

> the joy and the gratification that arise out of being in company with our peers, out of acting together and appearing in public, out of inserting ourselves into the world by word and deed, thus acquiring and sustaining our personal identity and beginning something entirely new.
>
> (Arendt, 1969: 263)

It is the pleasure of coming together and acting in concert that are peculiar to political life. Nevertheless, it is a zone of exclusion because not everyone has the possibility to compete, to be seen, and heard. As Benhabib states:

> The agonistic political space of the polis was only possible because large groups of human beings like women, slaves, laborers, non-citizen residents, and all non-Greeks were excluded from it and made possible through their 'labor' for the daily necessities of life that 'leisure for politics' which the few enjoyed; by contrast, the rise of the social was accompanied by the emancipation of these groups from the 'shadowy interior of the household' and by their entry into public life [...].
>
> (Benhabib, 1992: 91)

Arendt ignores that the equality of citizens is based on the oppression of those who are denied the status of being a citizen, those who take care of the necessities and work for the common world. In other words, the freedom of the political agents in her framework is built on the social, economic, and political exclusion and exploitation of others. Her approach towards action acknowledges difference and plurality, puts emphasis on the relational character of action, and highlights that we are always embedded in a net of human affairs. But Arendt seems to only regard the horizontal dimension of this net and not the hierarchical vertical lines cutting through. Arendt's political ideal place is not only a space of new beginnings and possibilities; it is also an impossible place that is inaccessible from the margins. This becomes even more problematic when we bear in mind that Arendt praises the polis not primarily as a city-state, but as an "organization of people as it arises out of acting and speaking together, and its true space lies between people living together for this purpose" (1958: 198). Ultimately, those who are not living for this purpose are excluded from political life – they are, as Arendt posits, dead to the world.

Arendt's Misrecognition

To unfold the doubts that over time have been raised regarding Arendt's model, we take a look at Moten's extensive study of Arendt's particular gaze. In his essay "Refuge, Refuse, Refrain" (2018), Moten focuses on the idea of intelligence associated with Arendt's work. He investigates the moral and political requirements that Arendt holds against subjects and their origin in "political theory's cool reason" (Moten, 2018: 88). Moten lays out the mechanics of "an antiblackness that infuses and animates Arendt's work" by taking a critical, yet interested look at her essay about the desegregation of Little Rock High School in 1957. Moten does not approach his object of study with the intention of 'proving her wrong'. Rather, his text offers a 'split gaze' that, on the one hand, reconstructs Arendt's motives and political beliefs and on the other hand, establishes black sociality as a sphere animated by its own (non-)political discourse. This gaze and Kathryn Gines' book on Arendt's more obvious judgemental failures regarding the case of segregation, help us to understand what it means that: "For Arendt, the parents of the Little Rock Nine were acting as members of society and not as political agents" (Allen, 2007: 319).

We have seen that, in order to be a political agent, one must act in a common world with others. As a structural model of this common world, Arendt imagines the Greek polis. In order to understand and compare Arendt's political stakes, Moten and Gines follow Elizabeth Eckford's "non-performance" in Little Rock. She is one of the nine black students being integrated at Little Rock Central High School and her picture in the newspaper gives way for Arendt's reflections to unfold.[4] Moten writes that Eckford's entering the school according to Arendt "has no honor in the polity" (Moten, 2018: 112) but, to Moten is an ante-political act in its "consent not to be a single being" (2018: 81). This means that Eckford's presence is being regarded as a non-political event in Arendt's eyes. It is non-political because what she does is not primarily on behalf of the republic's whole, but a cause of trouble with an unknown ending:

> The picture looked to me like a fantastic caricature of progressive education which, by abolishing the authority of adults, implicitly denies their responsibility for the world into which they have borne their children and refuses the duty of guiding them into it. Have we now come to the point where it is the children who are being asked to change or improve the world? And do we intend to have our political battles fought out in the school yards?
>
> (Arendt, 1959a: 50)

What she does not see is

> that beneath (her) notice they [the black parents of the nine children at Little Rock] (were) moved not as interested parties but as a party without interest concerned not with the transformation but with the liquidation of existing political institutions.
>
> (Moten, 2018: 100)

Eckford's motive when entering the high school is a non-performance, a "black refusal" because rather than establishing a "true and rightful position", which Arendt imagines to be the aim of her presence, Eckford is engaged in a "contract for destruction and rebuilding" (Moten, 2018: 101). In a sense, then, Arendt's reflections adequately capture the endangerment of the established social order – an event she fears so deeply. What Arendt is afraid of is the challenge Eckford presents to discrimination which is, according to Arendt, "as indispensable a social right as equality is a political right" (Arendt, 1959a: 51).

Accordingly, Moten reads Eckford's presence in the school as a disruption of undeterminable gravity. As a consequence of Arendt's conceptual anchoring in political theory, she denies the political quality and meaning of it. This denial does not only result in a protective attitude towards the institutions in question – in this case, the segregated school and neighbourhood – but is a symptom of her conviction that 'proper politics' do not entail disruptions of any sort (Moten, 2018: 101). Over the course of that argument – and Moten's reconstruction of it – the superior/elitist/exclusive character of politics within Arendt's thought is being revealed. Arendt seems to accuse the black parents of Little Rock of a betrayal of politics when they refused to accept the segregated order as a seemingly more peaceful arrangement than the tumultuous attempt at racial integration. Moten observes the boundaries of the polity clearly appearing here: the polity's citizens who "have honour" in the polity do not accept disruptive behaviour. Therefore, Elizabeth Eckford can only show herself as a person of the polity "who gains honor in the polity by accepting her exclusion from it" (Moten, 2018: 111). The political in the form of citizenship; which equals belonging to the polity of white society, is being weighed and protected against the social interest of black people (Moten, 2018: 110).[5]

Moten performs a 'political breakup' with Arendt when this exclusion reveals itself as a sacrifice that constitutes the social order:

> Technically, Arendt insists on the sacrifice of the noncitizen, the non-full citizen, the nonnaturalized, the unnatural who gains honor in the polity by accepting her exclusion from it. It's not that such sacrifice is imposed upon the one who is excluded; it is that exclusion constitutes the sacrificial activity of the citizen as he ritually makes of the excluded a sacrificial offering to the state.
>
> (Moten, 2018: 111)

Being ready to ignore the social interest that has been articulated by means of disruption means sacrificing the one who caused the disruption as a gesture that signifies trust in the state. The social interest with which Eckford enters the scene is accessibility to resources of education rather than securing the stability of the political order by any means. Moten states that Arendt's politics is one of anti-abolition. When abolition is the call for disorder against the reinforcement of the "already given segregation", when abolition means to expand "blackness as a refusal of a polity or community structured by refusal; blackness as a form of social thought in social life" then anti-abolitionism is the constraining order of white supremacy, "the American ideology", which makes Arendt "a representative figure of [...] American intellectuality" (Moten, 2018: 90). Beyond the breakup lies what Moten theorises together with Stefano Harney as "black sociality" or "fugitive sociality" (2013: 100) or "aesthetic sociality" as Laura Harris (2018) calls it. It is a space "outside of normative political agency" that crashes the "artificial boundary between sociality and publicness" that Arendt might not have erected in the first place but certainly restored (Moten, 2018: 100).

Danielle Allen argues in the context of Arendt's exclusive idea of the polity that "citizens should work to develop practices of and conversations about equity that will surround the polity's political institutions and provide a backdrop to its decisions" (Allen, 2007: 328). Her call for practices is meant to overcome the boundary that Arendt remains committed to. Practices, in this context, serve as an addendum that can soften the effects of state order at the margins of the institution. That is not an abolitionist approach, but it helps to see that the locus of practice is never at the centre. Moten's focus on Eckford emphasises the need for different concepts to approach her doing. Practice is a fitting term with various meanings. Saidiya Hartman uses it to understand how enslaved people used to organise their survival and therefore has a similar conceptual intention to Moten, who seeks to locate Eckford in her struggle.

What Is Practice?

Practice is a way for Hartman to describe actions that confront obstacles, hindrances, or prohibition. It also is a way of describing the usual, common mode of doing something against someone else's intention to prevent it. Also, how does it feel to act within constraints that have to be overcome over and over again without being able to overcome them for good? If political resistance very seldomly resembles the category of efficiency, it becomes a practice of resistance. Practice means repetition and improvement while it also provides for continuity: for Hartman, the everyday character of practice is key to understanding how survival is being organised under absolute domination. Since

> the tactics that comprise the everyday practices of the dominated have neither the means to secure a territory outside the space of domination nor the power to keep or maintain what is won in fleeting, surreptitious, and necessarily incomplete victories.
>
> (Hartman, 1997: 50)

Their re-enactment is what creates their efficiency. In terms of the political, practice describes activity not being regarded as political action as such because of the absence of authorship. The relevance for what is being called the political by Arendt though, is undeniable.[6] Hartman proposes to use practice as a means of thinking through resistance, powerless yet existing attempts to change the order of enslavement and acts of disobedience that leave no trace but are means to survive. In anticipation of the theoretical conditions of that discourse, she calls them "politics without a proper locus" (Hartman, 1997: 51).

Practice does not so much aim at change, it focuses on *redress*. Hartman says that redress "is undertaken with the acknowledgment that conditions will most likely remain the same" (1997: 51). And this knowledge of impossibility points towards the theoretical task of practice: it is a concept that helps to understand the politics of survival. In her study on *Undienlichkeit*, which translates as non-fungibility, Iris Därmann names and investigates such practices. In addition to Hartman's list of "work slowdowns, feigned illness, unlicensed travel, and destruction of property, theft, self-mutilation, dissimulation, physical confrontation with owners" (Hartman, 1997: 51), Därmann also describes the various techniques of self-harm and suicide (Därmann, 2020: 130ff). She contextualises acts of resistance as instances of initial creativity and articulates her bafflement about the possibility of the "power to reject and to self-withdraw" under conditions of enslavement and the constant threat of violent punishment (Därmann, 2020: 133). In order to extend the line of politics and practice, the discourse on resistance and enslavement must be understood as one about a "history that hurts" (Hartman, 1997: 51) and that still does. Enslavement is "the still-unfolding narrative of captivity, dispossession, and domination that engenders the black subject in the Americas" (Hartman, 1997: 51). While the becoming of such a subject reaches back to the Transatlantic Slave Trade and must be studied with that historical context in mind, we want to look at *practice* in our political present to understand where Arendt's concept of action confronts a certain limit.

When practice happens in a sphere conditioned by subjection, pain, and rightlessness, redress means finding ways to ameliorate these conditions. Not only is confrontative rebellion mostly pointless since efficient organisation is not possible, but it also constitutes a threat of punishment that might be far worse than enduring the usual. The ultimate goal of slipping away from violence can be achieved by different means. Many of them take into account the politics of performativity that constitute the order of enslavement (Hartman, 1997). Hartman analyses how the "demonstration of consent" (1997: 53) and enactments of the slave's happiness (1997: 17ff) were important tools to secure the violent racial order. The colonisers' reassurance that slaves were happy once they danced, made any claim for black pain and suffering impossible. Not only because the performance meant to prove the slave's enjoyment, but also because slaves merely served as vessels for any white projection and never were appointed feelings of their own (Hartman, 1997: 34). Rather, the slave's alleged happiness was proof of an extraordinary human condition that after all, showed whites to be bearable – it even allowed them to reassure themselves of their general empathy for the weak and suffering (Hartman, 1997: 35). The same is true for "representations of slave agency", which, instead of contributing to the autonomy they promised, helped to disguise the character of captivity. Hartman asks, "who is protected by such notions – the master or the slave?" (1997: 53) The emotional needs of the bourgeois subject – feeling like they allow for the slaves' happiness – constitute an aspect of emotional governance. This state of affective servitude for the enslaved means to exist in a twisted condition: performing happiness and contentment – against all odds and all pain – makes one important element to stay on relatively good terms with the master. It is a state of utmost vulnerability, described as "affectability" by

Denis Ferreira da Silva, who emphasises how this form of governance keeps the dominated subjects in their place (Ferreira da Silva, 2007, 2009). Practice, therefore, does not only entail all imaginable techniques of sabotage and unseen escape. Understanding and obeying affective and behavioural demands is part of it.

In addition to these particular complexities, assuming slave agency as such remains a difficult domain. Hartman notes that, considering all modes of performance that successfully preserve the slave-holder's power, "it is difficult to imagine a way in which the interpellation of the slave as subject enables forms of agency that do not reinscribe the terms of subjugation" (Hartman, 1997: 54). In order to live up to the challenge of liberating the enslaved subjects from the threat of reproducing the violence that constitutes their lives, Hartman seeks to interrupt the "subject as romance" (1997: 54). This expression refers to a bourgeois and sentimental idea of the white subject, which is depicted with a vivid inner emotional life that amongst other things implies moral feelings and concepts. Even though Arendt is not a thinker who has a general tendency for romanticisation, her political theory has a very strong idea of the capacity of singular, unique human subjects when they initiate actions in the mode of natality. Natality as a condition of political agency and universal quality of human beings is absolutely contrary to the lives of the enslaved, which Hartman wants to account for. This opposition could be further substantiated by Hartman's discussion of "natal alienation" (Patterson, 1982), which is the existential condition of the slave. It describes a state of enforced "kinlessness" which results in an absence of any lasting communal bonds or participation in a meaningful, consistent social sphere in which an Arendtian "beginning" would become recognisable in the first place. Arendt's view of subjective capacity assumes a sense of self that is irrelevant when the main orientation of an action is survival. Rather, all acts imaginable through original initiative action – natality – imply a kind of visibility. This visibility is problematic because it is either spatially impossible for enslaved people since they are removed from the public space or puts them in severe danger when they openly appear as disobedient and as deserving of punishment (Hartman, 1997: 55). Arendt, though, puts the "willingness to act and speak at all, to insert one's self into the world and begin a story of one's own" (Arendt, 1958: 186) at the centre of what it means to act. She says that no heroic qualities are necessary to succeed, but that "courage and even boldness are already present in leaving one's private hiding place and showing who one is, in disclosing and exposing one's self" (Arendt, 1958: 186). It is that courage and boldness of *showing oneself* that to the slaves did not promise freedom but would be answered with violence.

By suggesting practice, Hartman wonders what it means to *show* but not to *show oneself* as a means of ameliorating violence. The practice of performing blackness enacts this as blackness is "defined in terms of social relationality rather than identity" (Hartman, 1997: 56). Hartman's notion of practice emphasises the many acts which constantly reconfigure how subjects relate to their surround and each other. The focus on constant reorientation in a given field (De Certeau, 1984)[7] versus the individual unique initiation of singular acts distinguishes Hartman's concept of practice from Arendt's notion of action. Practice is always specific and it necessarily relates to racial violence when it is created under this condition.

How Can Action and Practice Still Relate to One Another?

Precisely, the conditions of violence and inequality inherent to the genealogies of practices seem to make practice incompatible with the narratives of freedom and equality Arendt is so adamant about in her conceptions of politics and action. For Arendt (2005), the locus of political action is the in-between of human beings but their relations cannot be structured by dominance and violence.

By holding up the division between the social and the political not only does Arendt dismiss what she regards as social questions, namely questions of discrimination and subjection, from the political agenda. She also dismisses whoever is denied equality and power due to discriminatory structures as a political agent. Hence, it appears that politics can only maintain its meaning by simultaneously denying it once concrete subjects try to claim their place under its conceptual realm. This inconsistency, which seems to be inscribed in Arendt's understanding of 'proper politics', can only be endured within an ideology that implicitly differentiates between legitimate and illegitimate agents. In this ideology, illegitimate political agents have to qualify as legitimate ones in order to be considered. Frantz Fanon sees this demand as a core element of "the bourgeois ideology that proclaims all men to be essentially equal" and only "manages to remain consistent with itself by urging the subhuman to rise to the level of Western humanity that it embodies" (Fanon, 2004: 109). Our critique towards the delineation of the political sphere and understanding of 'proper politics' is that it can only exist because access to it is made impossible for those who remain outside the sphere of public visibility. Political agency becomes an unequally distributed resource in this view and the problematic structures underlying the boundaries of the political realm remain unaddressed. Decolonising political agency then means addressing the structures and relations that make political agency and politics possible, that are produced by politics, and that are maintained by politics. This is pivotal since those structures and relations are themselves political as enabling and maintaining conditions of political agency and politics.

In order to think about these conditions, we need to elute the thinking of the political from the limitations of the public polity that is Arendt's model. This can be done through the concept of practice, which holds the promise of explaining the structures and dynamics politics is erected upon and of addressing those performances and events that an Arendtian concept of political action cannot consider. Through practice, the relations that first enable the in-between of people within the institutional structure of politics can be reflected. Accordingly, Arendt's relational concept of action is thus thought of in terms of the relations it implies but cannot explicate. This extends what we understand as the in-between of action, as well as the space of the political beyond the boundaries of the polity or other political institutions.

The point of practice, then, is to think about collective practices and to see how they qualify as political action without a proper agent. Moten's mantra of the "consent not to be a single being" that he adopted from Édouard Glissant's relational thought (Diawara et al., 2011) sketches agency in which agents are replaced by performance. This recalibrated form of political agency is not carried out by Arendt's heroes, but rather by a multiplicity of gestures and movements for which no single being holds the authorship. Thinking practice offers a perspective that shatters the concept of the political, lest it fall prey to what Arendt herself fears, namely, the inability to hear differences and to see and grasp reality (1970: 43). This is not to suddenly allow violence to enter politics, but to make visible the violence that is always already interwoven with the power to make politics. It is about seeing power and agency in collective practices and not just attesting to a few, thus dismissing any form of association outside the institution as disparate and a potential threat of violence. It is about not limiting ways of organising social life to the polity, but to regard that institution as one place among many to which one does not necessarily have to belong in order to be active.

Practice is vital when it comes in from the margins and sidesteps established ways of political intervention and institutionalised patterns of action. It is needed when these ways are foreclosed. This is also why we do not aim to simply add practice to the notion of agency or to integrate practices into the institution. Via practices questions of social interests can be articulated in the surroundings of the political boundaries they challenge. Therefore, it is crucial to not just adapt practices but to reckon the inherent performances of agency. Practice, after all, is part of the endeavour to decolonise agency by clarifying the concept's relation with institutional and societal structures conditioning the political field. The depoliticisation we aim for happens in the spirit of Moten and Harney's critique of politics as such that aims at institutional politics as a sphere of established interests operating through correcting the insufficient (Harney & Moten, 2013: 19). The forms through which these interests are being articulated are inherently anti-black.

The concept of agency is also somewhat enigmatic and blurry, something we want to take advantage of. This blurriness allows a theoretical embrace of minor and major gestures – regardless of whether they are a realisation of intention, a (non)performance or a collective endeavour – that otherwise remain unnoticed. Agency, we suggest, is not bound to form or position but a capacity of relation. We want agency which is not pre-regulated. This is what practice allows us to see. As such, it might become an altered model of how social interests are being negotiated. Practice takes into account performativity as one important capacity of the agent. We want agency to be as passive and minor as Hartman's gaze allows it to appear.

The Domain of the Senses

When we try to tie agency to a particular place, to a particular institution, we try to put our pin down on the proper locus. But every search for a proper locus, every pin moves on a map. As a point on a map, each locus, whether proper or improper, is connected to others and acquires its meaning through positionality, that is by the connecting lines and distance to other places. But lines are also overwritten, blurred or traversed. And is it at all meaningful and necessary to determine the proper locus between all the places that open up once we stop trying to focus and carve out that one place? The promise of clarity goes astray once we unfold the map. The locus can appear and be treated as property; but the ownership status becomes shattered, at least questionable, once we trench the soil it is built upon. Then dust swirls up, then smoke billows out. The points on our maps stand on sediments, they are based on genealogies that cannot be grasped with the clear gaze. Working with agency as a concept means establishing a cartography and genealogy of it. These systems of orientation are always specific. There is more to glance at and more to discover than one single stage where a few political subjects present themselves. Not only are there more stages – there are sites of practice, which exist with no spotlight and no seats for spectators. The cartography of agency unfolds to the margins and further, the map on which subjects position and navigate themselves shows the household, the marketplace, the ghettos. There are also things that are yet to be mapped and things that cannot be mapped, that refuse to be mapped, that are off-track, that are blurred. Our conceptual decisions function like spotlights on the map. We wander alongside existing lines and discover places hidden in paper creases. As we write, we draw new and shift old lines.

To look at a map, to travel down a genealogy is a sensual act. Depending on our condition and the time of the day, we see different things. There are times, the gaze unleashes and extends itself to foreign territory – even more so, when it appears as clear. In such a manner, Moten thinks that Arendt has collapsed the gaze of the mothers of Little Rock High School with her own conceptual one: "[...] the regulated imagination, the purified worldview, of the clear-eyed. The clear-eyed ask: How could one expose one's child to the (deadly) social cold, where everyone has the right (not) to associate with whomever they choose?" (Moten, 2018: 77). Arendt cuts through the visual and denies the cigarette smoke around her in order to ask the wrong question: "The question is not how to abolish discrimination, but how to keep it confined within the social sphere, where it is legitimate, and prevent its' trespassing on the political and the personal sphere, where it is destructive" (Moten, 2018: 51).

Her gaze moves with the intention to see, reveal, and clarify and therefore misses all the things that only exist in the fog, with the fog. Arendt's conceptual world is built in the light. It refuses to refer to a place that has none and simultaneously denies the existence of such a place. This has consequences for the kinds of politics that Arendt is interested in. The "clear, critical light of day"

(Harney & Moten, 2013: 20) makes transparency one of Arendt's main aesthetic motives.[8] Moten reminds us of Allen's "distinction between dark speech", that is, "a kind of unauthored and unauthorized speech whose origins cannot be traced", and "speech whose origins are visible, in the light of day, emerging from a place in the sun in which the rights and obligations of publicness, and the power to be seen, can be assumed" (Moten, 2018: 102).

Arendt craves "visible origins" of speech in order to fulfil the "need for institutions" (Harney & Moten, 2013: 20) that she feels. She won't accept the mystery of performative action that does not always know its cause of impact. The action, to her, is only complete with an agent that seems proper and properly visible. This is why "all institutions are political, and all politics is correctional" (Harney & Moten, 2013: 20). But practice breaks away from such a precondition. There is another way to see.

> I'm sorry if this is all a blur. I'm so used to my own astigmatism that maybe I can't even talk to anybody anymore. To make matters worse, I've never been able to keep my glasses clean. For the last forty-five years it's all been a blur [...] I think I'm seeing what I think I'm seeing, which makes me wonder if I'm seeing what I think. Hopefully it'll all be good in a minute, when I can stop talking to you and start talking with you.
>
> (Moten, 2017: 261)

When we cannot keep our glasses clean, we will have to talk ourselves through to find our way on the map. We ask neighbours and bystanders for the right way and most probably some of them will send us on a detour or point to a direction they think is right while wondering whether what they think is right. This mutual endeavour of pointing and guessing while meeting, talking, and discovering is what Moten and Harney often call *black study* and that is Arendt's greatest fear.[9] To overcome it, we have described an altered form of political agency as practice.

Notes

1 http://www.arendtcenter.it/en/tag/gunter-gaus/, all following quotes transcribed from the English subtitles of the video.

2 https://www.frieze.com/event/fred-moten-harmony-holiday-episode-1-quietness, all quotes transcribed from the video.

3 As Nullmeier aptly shows, Arendt is not concerned with victory or defeat in the political arena, and especially in political discourse, but with the will to show oneself and to appear. For Arendt, an agonal spirit leads to a double-sensed unfolding. On the one hand, it is about the unfolding of the individual, his individuality, in interaction with others. On the other hand, it is about the unfolding and discussion of different perspectives on an object. In favour of this unfolding, Arendt finally discards the agonal as a competition. Not the triumph of the best argument is decisive, but the possibility multi-perspectivity which can be won by common acting and speaking. "Arendt bids farewell to the agonal in favour of the will and the ability to change perspective and adopt perspective as the result of a public discourse of many" (Nullmeier, 2000: 174).

4 In her "Reply to Critics", Arendt recalls her initial impulse to write about Little Rock. Next to the worry about the federal court order, her motivation for writing about Little Rock seems to start off with an affective reading of the photograph:

> The point of departure of my reflections was a picture in the newspapers, showing a Negro girl in her way home from a newly integrated school; she was persecuted by a mob of white children, protected by a white friend of her father and her face bore eloquent witness to the obvious fact that she not precisely happy.
>
> (Arendt, 1959b: 179)

Gines revisits Arendt's description and remarks that she was mistaken about the people she identified in the picture as well as their relationship (Gines, 2014). Gines criticises Arendt's presumptuous reading of the picture and sees a problematic tendency at work here. Often, Arendt's projections lead to misjudgements and oversimplifications about the Black community's stakes in school and integration, Black parenting, and Black people's perspectives on successful social change (Gines, 2014: 20ff).

5 This force, and the exposure it demands, remain unspeakable for Arendt. Her citizenship – her allegiance to a given politics, its constitution and its institutions – inclines her not only to reject whatever disrupts these institutions but to characterize such disruption as unnaturally anti- and antepolitical.

(Moten, 2018: 101)

6 In her book *A Feminist Theory of Refusal* (2021), Honig brings Arendt and Hartman into conversation. She discusses Hartman's use of fabulation, a form of counter-narrating, as a method of refusal, and the practices of refusal performed by Hartman's wayward women (Hartman, 2019). Honig shows how Hartman's fabulation and wayward women challenge Arendt's distinction of the social and political as well as the form of collectivity that Arendt's understanding of political action presupposes. Honig also aptly describes moments and places where Hartman and Arendt meet, for example, in the agon or the archive.

7 Hartman refers to Michel de Certeau's *The Practice of Everyday Life* in order to conceptualise practice as "'a way of operating' defined by 'the nonautonomy of [a] field of action'" (De Certeau, 1984: 21 as cited in Hartman, 1997:50).

8 Édouard Glissant (2010) works on the opposition of transparency and opacity and explains the distinction we aim at here in a more fundamental way.

9 In *On Violence*, Arendt writes about violence in the civil rights movement and the student movement in West Germany. She derogatorily refers to Black Studies classes as "soul courses": "No doubt, 'violence pays,' but the trouble is that it pays indiscriminately, for 'soul courses' and instruction in Swahili as well as for real reforms" (Arendt, 1970: 79–80). Arendt also writes:

> In America, the student movement has been seriously radicalized wherever police and police brutality intervened in essentially nonviolent demonstrations: occupations of administration buildings, sit-ins, et cetera. Serious violence entered the scene only with the appearance of the Black Power movement on the campuses. Negro students, the majority of them admitted without academic qualification, regarded and organized themselves as an interest group, the representatives of the black community. Their interest was to lower academic standard.
>
> (1970: 18)

In the German translation, Arendt adds that because it might not be enough to lower academic standards, so Black Students could keep up, they demand their "own 'study area', the so-called 'Black Studies'" (1970: 22). Not only is Arendt dismissive of Black Studies and thereby discards Black Culture, History, and Knowledge. She also denies the capabilities of Black Students to adapt to the specific intellectual demands of white academia. In this manner, she accounts for deeply racist assumptions and disguises them as neutral statements of social progress. Her words, to put it differently, contribute to the continuation of exclusion and oppression.

References

Allen, D. (2007). Law's necessary forcefulness: Ralph Ellison vs. Hannah Arendt on the Battle of Little Rock. In A. Laden & D. Owen (Eds.), *Multiculturalism and Political Theory* (pp. 315–349). Cambridge, UK: Cambridge University Press.

Arendt, H. (1958). *The Human Condition*. Chicago: Chicago University Press.

Arendt, H. (1959a). Reflections on Little Rock. *Dissent*, *1*(6), 45–56.

Arendt, H. (1959b). A Reply to Critics. *Dissent*, *2*(6), 179–181.

Arendt, H. (1969). *Between Past and Future. Eight Exercises in Political Thought*. New York: The Viking Press.

Arendt, H. (1970). *On Violence*. San Diego, New York and London: HBJ Publishers.

Arendt, H. (2005). *The Promise of Politics*. Ed. by J. Kohn. New York: Schocken Books.

Benhabib, S. (1992). *Situating The Self. Gender, Community and Postmodernism in Contemporary Ethics*. Cambridge, UK and Malden, MA: Polity Press.

Därmann, I. (2020). *Undiendlichkeit - Gewaltgeschichte Und Politische Philosophie*. Berlin: Matthes und Seitz.

De Certeau, M. (1984). *The Practice of Everyday Life*. Trans. by S. Rendall. Berkeley, Los Angeles and London: University of California Press.

Diawara, M., Glissant, É., & Winks, C. (2011). Édouard Glissant in Conversation with Manthia Diawara. *NKA Journal of Contemporary African Art*, *28*(1), 4–19.

Fanon, F. (2004). *The Wretched of the Earth*. Trans. by Richard Philcox. New York: Grove Press.

Ferreira da Silva, D. (2007). *Toward a Global Idea of Race*. Minneapolis: University of Minnesota Press.

Ferreira da Silva, D. (2009). No-Bodies: Law, Raciality and Violence. *Griffith Law Review*, *18*(2), 212–236. https://doi.org/10.1080/10383441.2009.10854638

Gaus, G., & Arendt, H. (1964). *Zur Person. Hannah Arendt im Gespräch mit Günter Gaus* [Video]. http://www.arendtcenter.it/en/tag/gunter-gaus/

Gines, K. T. (2014). *Hannah Arendt and the Negro Question*. Bloomington: Indiana University Press.

Glissant, É. (2010). *Poetics of Relation*. Ann Arbor: University of Michigan Press.

Harney, S., & Moten, F. (2013). *The Undercommons: Fugitive Planning & Black Study*. Wivenhoe: Minor Compositions.

Harris, L. (2018). *Experiments in Exile: C.L.R. James, Hélio Oiticica, and the Aesthetic Sociality of Blackness*. New York: Fordham University Press.

Hartman, S. V. (1997). *Scenes of Subjection: Terror, Slavery, and Self-Making in Nineteenth Century America*. Oxford: Oxford University Press.

Hartman, S. V. (2019). *Wayward Lives, Beautiful Experiments. Intimate Histories of Social Upheaval*. New York: W. W. Norton & Company.

Honig, B. (1992). Toward an Agonistic Feminism. Hannah Arendt and the Politics of Identity. In J. Butler & J. W. Scott (Eds.), *Feminists Theorize the Political*. New York: Routledge, 215–235.

Honig, B. (1993). *Political Theory and the Displacement of Politics*. Ithaca and London: Cornell University Press.

Honig, B. (2021). *A Feminist Theory of Refusal*. Cambridge, MA: Harvard University Press.

Marchart, O. (2007). *Post-Foundational Political Thought. Political Difference in Nancy, Lefort, Badiou and Laclau*. Edinburgh: Edinburgh University Press.
Moten, F. (2017). *Black and Blur*. Consent Not to Be a Single Being: V. [1]. Durham: Duke University Press.
Moten, F. (2018). *The Universal Machine*. Consent Not to Be a Single Being: V. [3]. Durham: Duke University Press, 2018.
Moten, F., & Holiday, H. (2020). *Episode 1: Quietness* [Video]. *Frieze*. https://www.frieze.com/event/fred-moten-harmony-holiday-episode-1-quietness
Nullmeier, F. (2000). *Politische Theorie des Sozialstaats*. Frankfurt and New York: Campus Verlag.
Patterson, O. (1982). *Slavery and Social Death*. Cambridge, MA: Harvard University Press.

5 Enfleshed Political Violences

Rethinking Sexual Violence from a Decolonial Critique to the Political Construction of the Body as Flesh

Cecilia Cienfuegos Martínez

Sexual violence is always political. This is a productive statement because this form of violence is usually taken away from a political reflection. In fact, if anything, sexual violence seems to carry with it some sort of de-politicising effect. It locks its victims in the personal and intimate realms, which in a modern/colonial framework means the non-political. As a consequence, sexual violence is usually recognised as a form of political violence only when it is instrumentalised within a conflict previously defined as political; when it is used, for instance, as a war weapon, as a form of torture or as a tool for the destruction or total demoralisation of a community. In any case, the qualification of sexual violence as political depends on its context and, usually, on the explicit intentions of its perpetrators. The central hypothesis of this chapter is that sexual violence is always a form of political violence of some sort, regardless of its particular use, no matter the intention of the perpetrator/s. Even more, it will be argued that sexual violence is one of the privileged forms of violence in the imposition of gendered and racialised orders of power in the "modern/colonial gender system" (Lugones, 2008). A subsequent argument is that in order to analyse sexual violence it is important to keep in mind that this form of violence does not mean the imposition of *the masculine power* over *the feminised body*, but a productive force in the construction of gendered and racialised distinctions that are not susceptible to universalisation under the paradigm of sexual difference (Alcoff, 2018).

The first half of the chapter argues that the work of decolonial feminism helps to understand to what extent the imposition of the modern distinction of public/private as political/non-political becomes pivotal in the construction of the colonial order of bodies and power. This distinction is at the core of the colonial imposition of the gender system (Lugones, 2008), and relegates the intimate to the apolitical (Segato, 2016a). While, at the same time, it allows for the systematic instrumentalisation of sexuality and, paradigmatically, of sexual violence, in the imposition of gendered and racialised distinctions as pre-political. The second half of the chapter introduces Hortense Spillers' (1987) and Alexander Weheliye's (2014) theories on the notion of *flesh* to illustrate the historical continuities that define the violent onto-material dichotomisation of bodies and violence brought by the historical experiences of slavery and

DOI: 10.4324/9781003293460-8

colonisation. These dichotomies cannot be overlooked in the thinking of contemporary political violence, since gendered and racialised frameworks mark different bodies and make them subjective to forms of violence that would be unthinkable in bodies framed by different variables under the modern/colonial order of power. Ultimately, the notion of flesh will be enriched by Elsa Dorlin's (2018) theory on the "defensive apparatus" of power that comes to define coloniality as a hierarchised access to the possibility of violence and self-defence. Dorlin's theory helps to understand to what extent the different archives of cruelty define different frameworks of what is thinkable, and different consequences regarding who can be a legitimate agent and who is condemned to be the object of the most *spectacularised* (Hartman, 1997) forms of violence.

To summarise, the ultimate proposal of this chapter is to interrogate the encounter between decolonial theory, black studies, and feminist critiques in the study of sexual violence, with the goal of defining some of the political limits in the thinking of this form of violence, as well as the political implications of these limitations. Some of the questions that articulate this chapter are: First, what does it mean to think of sexual violence as political and what are the mechanisms involved in its depoliticisation in the modern/colonial gender system? Second, how does sexual violence participate in the 'making' of the colonial, gendered, and racialised orders of bodies and flesh? And, finally, how do these considerations on the differential orders of bodies and violence transform the very definition of the political? The approximation to these questions helps to envision the complexities of the matter, as well as conceptually delimit some of the most relevant aspects of what comes into play in the thinking of sexual violence.

Coloniality of Gender and the Mark of the Human

Decolonial feminist theory is diverse but presents different alternatives to some central problems. On the one hand, decolonial theorists develop a critique aimed at a form of hegemonic feminism that presupposes Western solutions as universal, from a position of moral superiority regarding non-white and non-Western women in need of civilisation/Westernisation (Khader, 2018). And, on the other hand, they agree on the need to include gender at the core of the analysis of the modern/colonial capitalist system. This is so because, as María Lugones (2008) argues, colonial modernity depends on a particular organisation of gender and sexuality. Gender, in this framework, is thus a central epistemic and political category, "capable of illuminating all other aspects of the transformation imposed on the lives of communities as they were captured by the new modern colonial order" (Segato, 2016a: 111). Hence, for these theorists, to attend to the transformations in the orders of gender and violence becomes a methodological starting point to address broader dynamics in the capitalist world system of power.

However, there is a central discussion among decolonial feminist theorists regarding the extent to which it is possible to talk about gender before colonisation. In this dispute, Oyèrónkẹ Oyěwùmí presents one of the most compelling arguments. Taking the analysis of precolonial Yoruba society as a central example, Oyěwùmí (1997) argues that gender is strictly a colonial imposition, which cannot be thought outside of Western modernity. According to her study:

> The emergence of women as an identifiable category, defined by their anatomy and subordinated to men in all situations, resulted, in part, from the imposition of a patriarchal colonial state. For females, colonization was a twofold process of racial inferiorization and gender subordination.
>
> (Oyewùmí, 1997: 124)

Hence, on the basis of Oyěwùmí, Lugones (2008) critiques Aníbal Quijano's notion of the *coloniality of power* (2000) for not going far enough in his understanding of gender.

According to Lugones, whereas Quijano includes gender as a key variable in the shaping of the modern/colonial order of power, he fails to envision the fictional dimension and historical contingency of gender itself. She ultimately argues that, while presenting an understanding of race as a central fictional creation of the colonial expansion, Quijano reproduces a Westernised understanding of gender, defined by biological dimorphism, patriarchal organisation of gender relations, and heteronormativity (Lugones, 2008: 2). In developing her notion of the coloniality of gender, Lugones argues these three aspects need to be understood contextually as the result of a historical development of gender and sexual relations that only applies to Western modernity – while, in the colonies, the same categories are established as a horizon of the (im)possible, marking racialised bodies as ontologically degendered.

Hence, critiquing Quijano, Lugones maintains that there are two sides of gender: a "light side" of the modern/colonial gender system (reserved for white subjects), and a "dark side" of this order of gender, which operates in the colonies: "The light side constructs gender and gender relations hegemonically. It only orders the lives of white bourgeois men and women, and it constitutes the modern/colonial meaning of 'men' and 'women'" (Lugones, 2008: 15). Meanwhile, the "dark/invisible" side of gender applies to colonised bodies by exclusion: marking them as bodies that do not fit into the categories of 'men' and 'women'. According to Lugones, the only way to adequately address the dark/invisible side of gender is to attend to the experience of women of colour from an intersectional paradigm (Crenshaw, 2017). That is, to think beyond the categorial logic enforced by Western modernity, which imposes the categories of gender, race, sexuality, or class as if it were possible to think of any of them separately. The problem with this logic is that these "categories have been understood as homogenous and as picking out the dominant in the group as the norm" (Lugones, 2008: 4), thus rendering invisible the lives and experiences of those who do not fit the hegemonic version of each position.

Quijano's fault would be to take the visible side for granted as an ontological and non-historical truth of gender – thus abstracting from the fact that gender itself, as race, is a key political fiction, instrumentalised in the imposition of the colonial order of power. Indeed, if anything can be recognised in analysing the genealogy of the imposition of the colonial power, it is an instrumentalisation of the normative structures of gender as a mark of the human from which colonised subjectivities are systematically expelled. That is, albeit a hegemonic definition of gender is universalised and naturalised through European colonial expansion, this universalisation should not be understood literally. What Lugones points out is that, while gender categories come to be imposed throughout the world with colonisation, the definition of colonised subjectivities is precisely marked by the expulsion from this order. "The behaviors of the colonised and their personalities/souls were judged as bestial and thus non-gendered, promiscuous, grotesquely sexual, and sinful" (Lugones, 2010: 743). As a result, in the modern/colonial order of gender "colonised people became males and females. Males became non-human-as-not-men, and colonised females became not-human-as-not-women" (Lugones, 2010: 744). Hence, just like race, gender represents a key instrument in the hierarchisation of the world's populations in relation to the central dichotomy imposed by colonial modernity: that of the human and the non-human (Maldonado-Torres, 2007). And this dehumanisation as a process of de-gendering that defines the "dark side" of the gender order "was and is thoroughly violent" (Lugones, 2008: 16). The third and fourth sections of this chapter will devote special attention to these violent interventions. But first, the next section analyses the modern/colonial process of depoliticisation of the private and intimate spheres as a crucial step in the depoliticisation of these violent constructions.

The Depoliticisation of the Private

With regard to the aforementioned discussion, Segato argues that both Lugones and Oyěwùmí overlook the universality of patriarchal oppression. She maintains that there is sufficient evidence worldwide to argue that gender works as a political category long before colonisation, demarcating differences in status and power between men and women (Segato, 2016a: 112). However, while arguing that gender existed before colonisation, Segato concurs that there were major differences with the modern organisation of gender:

> [I]f it is possible to say that there was always hierarchy and gender relations as relations of unequal power and prestige, with the colonial state intervention and the entry into the order of colonial/modernity that oppressive distance is aggravated and magnified.
>
> (Segato, 2016a: 123)

According to her analysis, the colonial invasion translates into a new super-hierarchical world, which is accompanied by three main consequences: the binarisation of dualisms; the hyperinflation and universalisation of the public sphere; and the emasculation of colonised men in the extra-communitarian world (Segato, 2016a: 113). These three main transformations are key to understanding the colonial orders of racialisation and genderisation in terms of violence. Let us first explore the first two, which are intimately related. The latter will be addressed below, in relation to a further development of the complex and paradoxical constructions of colonised gendered subjectivities in their relation to violence.

With the idea of the binarisation or dichotomisation of duality, Segato means that what functions as a duality in terms of gender in what she names the "village-world" (in which both sides, although hierarchical, have ontological and political plenitude), is transformed into a dichotomy in the modern/colonial world. This dichotomisation means that one side (the masculine) comes to represent the universal, while the second term (the feminine) becomes residual: the negative side of the norm. And in political terms, the problem is that, with modernity, only those who inhabit the universal are considered political subjects. In other words, this ontological dichotomisation turns into a political one, which results in the depoliticisation of the private sphere. By way of contrast, analysing examples from the Amazonian, Chaco, and Andean worlds, Segato identifies that, although two spaces (public and private) can be distinguished and are organised in terms of gender in these societies, both spaces are ontological totalities. Moreover, both participate in the political debate. That is to say:

> there is no monopoly of politics by the public space and its activities, as in the modern colonial world. On the contrary, the domestic space is endowed with politicization, because it is of obligatory consultation and because in it the corporate group of women is articulated as a political front.
>
> (Segato, 2016a: 117)

The confinement of the political to the public space is a hallmark of the modern social-sexual contract (Pateman, 1988). And what inhabits the negative and private realm, as feminist theory has thoroughly theorised, is expelled from this contract. Here, therefore, Segato ultimately agrees with Lugones and argues that the imposition of the modern/colonial system brings with it an unprecedented order of gender, defined by "the abduction of all politics" by the realm of the masculine: the "privatization of the domestic space, its otrification, marginalization and expropriation of all that was political in it" (Segato, 2016a: 116). Thus, what coloniality imposed globally is the depoliticisation of the private sphere, as well as the idea of the intimate as that space that must be safeguarded from political interference. In this process, the intimate is excluded from the public gaze and, therefore, abandoned to its own logics as if these were not of community interest. The depoliticisation of sexual violence in colonial modernity cannot be understood in isolation from these exclusions of the intimate and the private from the political, key in what Segato names the *minoritisation of the feminine* (2016b) in modernity/coloniality.

Hence, colonisation brought with it a radical decrease in women's political power (Gautier, 2005; Oyěwùmí, 1997), but also a rupture of community ties (Federici, 2004) and the extreme victimisation of those confined to the private sphere (defined as non-political and thus excluded from the vigilance of the communitarian eye). But this victimisation does not only play out in terms of gender. Gender, race, and the very categories of coloniser and colonised need to be expanded in their definition beyond the modern dichotomic categorial logic. This is so because, in Oyěwùmí's terms:

> in the colonial situation, there was a hierarchy of four, not two, categories. Beginning at the top, these were: men (European), women (European), native ([colonised] men), and Other ([colonised] women). Native women occupied the residual and unspecified category of the Other.
>
> (1997: 122)

In other words, with conquest and colonisation comes the institution of the social-sexual-racial contract (Pateman & Mills, 2007), and neither the feminine nor the colonised should be understood as homogeneous categories.

As will be exposed below, this depoliticisation of the private has extreme consequences in terms of violence; with a private sphere defined by the depoliticisation of intimate and gender-based violence, and a construction of racialised hierarchies that stands in the way of identifying and condemning these forms of violence on racialised individuals. That is, being able to analyse *these violent constructions requires an intersectional and decolonial approach*. In this sense, despite their divergences, the idea that the violent construction of a racialised order of gender is a fundamental device in the ongoing imposition of coloniality is maintained on both sides of the debate. This is the idea that interests me – as well as the extent to which sexual violence participates in this ongoing political construction. The following section focuses on the notion of *flesh* (Spillers, 1987; Weheliye, 2014) as a key theoretical contribution to the thinking of the violent construction of racialised subjectivities, in order to ultimately introduce the productive capacity of violence in the creation of the modern/colonial order of bodies and power.

Theorising the *Flesh*

In *Habeas Viscus* (2014), Alexander G. Weheliye starts his work by reflecting on the encounter with the notions of *bare life* (Agamben, 1998) and *biopolitics* (Foucault, 1990), but critically points out that both Agamben and Foucault misconstrue "how profoundly race and racism shape the modern idea of the human [...] allowing bare life and biopolitics discourse to imagine an indivisible biological substance anterior to racialization" (Weheliye, 2014: 4) and, it should be added here, following Lugones, to genderisation and sexualisation. Weheliye concretely critiques Agamben's notion of bare life because he argues that, with it, Agamben "locates the political digestion of *zoe* in a generalised, quasi-ontological 'zone of indistinction' in which the categories that segregate bare life and other modes of life become obsolete" (Weheliye, 2014: 34). That is, the problem is that Agamben presupposes a sort of 'original substance' of life that seems to transcend the divisions of race, class, gender, nationality, or sexuality that articulate social and political life.

It is in relation to this critique that Weheliye reclaims Spillers' distinction between the notions of *body* and *flesh*, which she defines in her article "Mama's Baby, Papa's Maybe" (1987) as "the central [distinction] between captive and liberated subject-positions" (Spillers, 1987: 67). For Weheliye, such contribution helps to correct Agamben's and Foucault's approaches "by highlighting the embodiment of those banished to the zone of indistinction and by showing how bare life is transmitted historically so as to become affixed to certain bodies" (Weheliye, 2014: 38). This notion thus articulates a more nuanced starting point towards any attempt to study oppression. In this regard, Weheliye uses the expression *Habeas Viscus* ("You shall have the flesh"), on the one hand,

> to signal how violent political domination activates a fleshly surplus that simultaneously sustains and disfigures said brutality, and, on the other hand, to reclaim the atrocity of flesh as a pivotal arena for the politics emanating from different traditions of the oppressed.
>
> (Weheliye, 2014: 2)

That is, this concept helps to account for a different notion of politics: one inscribed in the operations of the quotidian violence that shape and produce oppressed subjectivities.

Thus, it is a matter of paying attention both to how power is predicated on violence, and to how the forced embodiment of brutality functions to preclude the operations of power. It is in this sense that flesh is 'outside' of discourse and 'outside' of culture – or, more accurately, outside of cultural and political concerns. In Spillers' words: "Before the 'body' there is the 'flesh,' that zero degree of social conceptualisation that does not escape concealment under the brush of discourse, or the reflexes of iconography" (Spillers, 1987: 67). But it would be a mistake to think that the flesh 'precedes' the body in any chronological way. Quite the contrary, the captive body is the result of an elaborate process of enfleshment. The flesh is, quite literally, the political product of violence. In other words, the flesh is not 'more natural' than the body. It does not 'precede' the cultural or political spheres, but it is constructed precisely through the violent expulsion from these realms. In Spillers' words, the appearance of a body as flesh is the result of "the calculated work of iron, whips, chains, knives, the canine patrol, the bullet" (Spillers, 1987: 67) – and, of course, of sexual violence. That is, it requires an extremely violent intervention, where the manufacturing of the flesh comes with the literal and symbolic tearing apart of the body.

The theft of the body that comes with colonialism and enslavement is therefore defined by "a willful and violent (and unimaginable from this distance) severing of the captive body from its motive will, its active desire" (Spillers,

1987: 67). However, as Spillers illustrates, the possibility of slavery was based on constant violent interventions over black bodies which articulated a horizon of the possible with regard to the violation of those bodies 'marked' as flesh, transcending their punctual existence in history. Spillers develops this idea by referring to what she names the "hieroglyphics of the flesh", a notion that accounts for the historical continuities that conform racialised embodiments. She uses this notion to illustrate the incarnated (collective) memories of different forms of past violence that still come to define black bodies in the present. In Weheliye's words:

> What Spillers refers to as the "hieroglyphics of the flesh" created by these instruments [of violence] is transmitted to the succeeding generations of black subjects who have been "liberated" and granted body in the aftermath of the jure enslavement. The hieroglyphics of the flesh do not vanish once affixed to proper personhood (the body); rather they endure as a pesky potential vital to the manoeuvrings of "cultural seeing by skin color".
>
> (Weheliye, 2014: 39–40)

That is to say, the exposure to violence and the very exposure of the violated enfleshed body are fundamental tools for the internalisation of submissiveness in the realm of the flesh. But it also constructs an ontological distance towards the sufferings of enfleshed subjectivities. In this ongoing process, the visuality of these forms of violence particularly 'defined' for the exploitation and carnage of the captive body is also central to the black body becoming flesh. Spillers, once again, verbalises these processes since, according to Weheliye, she "interrogates the visual, fleshly distinctions that comprise the nexus of racialization and/as bare life" (Weheliye, 2014, 38). To do so she introduces the notion of *pornotroping*, by which she reflects on the visuality of enfleshment. In Spiller's words: "as a category of 'otherness,' the captive body translates into a potential for pornotroping and embodies sheer physical powerlessness that slides into a more general 'powerlessness', resonating through various centres of human and social meanings" (Spillers, 1987: 67).

It is from a similar vein that Saidiya V. Hartman (1997) frames her concern about "the spectacular nature of black suffering and, conversely, the dissimulation of suffering through spectacle" (Hartman, 1997: 22). While the quotidian violent formation of genderisation and racialisation remains naturalised, the spectacle of extreme violence over black bodies systematically reproduces the 'logic of subversion', according to which enfleshed suffering is attended to only as a spectacle, and only in its most extreme expressions. That is, Hartman theorises the problematic *spectacularisation* of violence that precludes the possibility of any genuine form of *empathy* towards black suffering. In a very similar line of argument, Weheliye articulates a challenging question, which has relevant consequences for the thinking of the present: "What do distinctive manifestations of black suffering at the hands of political brutalization in slavery and beyond tell us about the general function of politics and/as suffering?" (Weheliye, 2008: 66).

The captive body is the body whose suffering is not just visible but actively exposed, and the imposition of 'seeing', in the form of enforced witnessing of violence over racialised bodies, appears as a fundamental means for internalising the extreme vulnerability and dispensability that marks enfleshed subjectivities. As argued by Hartman, the extreme and reiterative exhibition of the black body in pain, and the forced witnessing of this suffering, defines a very particular relation to violence for the captive body. One that constructs a conscience of absolute disposability and comes to normalise a visuality of the ripped black body smashed by forms of violence that would be inconceivable for a white body. This is what explains that "rather than entering a clearing zone of indistinction, we are thrown into the vortex of hierarchical indicators: racializing assemblages" (Weheliye, 2014: 40). And these "racializing assemblages" refer to the translation of the lacerations and markings of the captive body by systemic and political violence. Markings that, following Spillers, do not just 'disappear' with the abolition of the systems themselves (slavery or colonisation). Instead, they endure beyond abolition with racialised bodies remaining enclosed in the paradigm of the flesh that defines modernity/coloniality.

(In)Defensible: Sexual Violence and the Imperial Economy of Violence

As discussed so far, the organisation of bodies and violence constitutes a central distinction in the construction of modernity/coloniality. However, in this process, not only the enfleshed subjects get transformed. Violence itself becomes adjusted to the necessities of this process, with forms of violence literally reserved for the destruction of the captive body. And the very sovereign subject appears as the product of this distinction. In fact, as Aimé Césaire stated in his *Discourse on Colonization* (2000), the possibility of the coloniser's subjectivity does not only require the dehumanisation and violation of the Other but, most importantly, the complete brutalisation of the self. It demands "to degrade him [sic], to awake him to buried instincts, to covetousness, violence, race hatred, and moral relativism" (Césaire, 2000: 35). That is, in order to make domination even possible, the very dominant subject is also transformed and produced in the construction of the distinction between body and flesh, as a paradigmatic agent of violence.

Building on the same attention to the relevance of regarding violence in the conformation of modern subjectivity, Dorlin (2018) argues that to focus on the differential access to violence (and, in her analysis, the possibility of *self-defence*) points to a portrait of the construction of the modern subject "but in negative relief" (Dorlin, 2018: 12). What comes to be established with modernity/coloniality is the capital distinction between bodies who have access to legitimate defence and the right to be protected by the state; and bodies who are radically dispossessed and ontologically disarmed. That is, the fleshly surplus of the modern/colonial project is defined, by contrast to 'full subjectivity', by a delegitimised access to violence: "those whose capacity for self-defense needed to be either diminished or annihilated, corrupted or delegitimated, and whose defensive bodies were exposed to the risk of death" (Dorlin, 2018: 21). Hence, while the hegemonic narrative of the construction of the modern state tells a story of the progressive monopolistic management of violence by states, centralising the colonial origins of the captive body tells a very different story. In this regard, Dorlin identifies the creation of what she names an *imperial economy of violence* "that paradoxically defends individuals who have always already been recognised as legitimate defenders of themselves" (Dorlin, 2018: 29). What can thus be identified in the global modern colonial system is a new form of power based on the distribution of the possibility of violence.

In this new paradigm, a defence dialectic is imposed according to which the possibility to defend oneself is a privilege of the dominant minority. However:

> This dividing line is never solely about distinguishing threatening/aggressive bodies from defensive bodies. Rather, it separates those who are agents (agents of their own defence) from those whose power of action has an entirely negative form, in the sense that they can only ever be agents of 'pure' violence.
>
> (Dorlin, 2018: 26–7)

And in this order of power, which Dorlin defines as the creation of a "defensive apparatus", the racial instrumentalisation of the "defence of *our* women" discourse occupies a pivotal position. Over this rhetoric, of course, sexual violence holds a central role and gets systematically perverted under the paradigm of white supremacy (see, as paradigmatic examples: the analysis of the phenomenon of lynching and the myth of the black rapist as developed by Wells, 1892; Davis, 1981; or Dorlin, 2018). In this context, however, the important relation is not that between the victim and the alleged offender, but between the subject who defends and, on the one hand, the one who is defended (white women, defined as ontologically vulnerable, disarmed, and in need of protection), and, on the other hand, the one from whom she supposedly needs to be protected (the racialised man, ontologically threatening and, ultimately, killable). A fourth figure, systematically taken out of the picture, is the racialised woman – who, in a correlative process to the brutalisation of racialised men, becomes ontologically rapeable and not worthy of protection.

Now, here comes an important set of paradoxes. First of all, the position of white women in this rhetoric is indeed ambiguous. In Wendy Brown's terms:

> Protection codes are thus key technologies in regulating privileged women as well as in intensifying the vulnerability and degradation of those on the unprotected side of the constructed divide between light and dark, wives and prostitutes, good girls and bad ones.
>
> (Brown, 1995: 170)

Hence, Dorlin defines white women as "impure subjects", as they are "both object and subject of the politics of *race defence*" (Dorlin, 2018: 153). Meanwhile, racialised subjectivities have to negotiate in a very paradoxical terrain regarding violence and sexuality. On the one hand, colonised femininity is defined as always sexually accessible (rapeable) and hypersexualised, but at the same time is often thought of as unattractive and/or aggressive. In fact, even though systematically exposed to sexual violence, rape often appears as inconceivable in the case of racialised women: either because they are expropriated from the possibility of consent or resistance (as in the case of enslaved women), and/or because their sexuality is thought of as lascivious and excessive (Hartman, 1997: 81).

On the other hand, colonised masculinity is understood as hypersexual and threatening, while simultaneously, as previously stated by Segato (2016a: 113), another central feature of the construction of the colonised is the emasculation of colonised men in the extra-communitarian world. Moreover, in Maldonado-Torres' account: "Black bodies are seen as excessively violent and erotic, as well as the legitimate recipients of excessive violence, erotic and otherwise. 'Killability' and 'rapeability' are part of their essence – understood in a phenomenological way" (Maldonado-Torres, 2007: 255–6). Hence, while coloniality imposes primarily an order of killability and rapeability in which women of colour are positioned as primary targets, in this order of violence "[m]en of color are feminised and become for the ego conquiro fundamentally penetrable subjects" (Maldonado-Torres, 2007: 248). The intervention of the threat of rape in this paradoxical construction is crucial, and it applies in equally conflicting terms to male and female racialised subjects – thus transcending the heteronormative script imposed by the hegemonic narrative of sexual violence (Hartman, 1997: 80). These ambiguous relations between violence, race, and sexuality are mediated by contextual processes of racialisation that cannot be universalised. But what remains is the possibility of ambiguity in the service of brutalisation, which is a constant in the enfleshed formation of the captive body. Consequently, both raping and being potentially rapeable are intrinsic possibilities of colonised subjects. This is part of what Maldonado-Torres (2008) refers to as the *non-ethics of war* as the paradigmatic state of colonised existence.

Nonetheless, according to Segato, the emasculation of colonised men has particularly brutal consequences for the lives of colonised women, whose bodies are turned into territories at war: "as they became progressively more vulnerable to men's violence, in turn potentiated by the stress caused by the pressure on them from the outside world" (Segato, 2016a: 116). As a matter of fact, the imposition of the modern/colonial gender order becomes a key instrument in the rupture of community ties within the colonial world, and sexual violence is a key element in this process: "At times, it was the 'Indian' men themselves who delivered their female kin to the priests or *encomenderos* in exchange for some economic reward or a public post" (Federici, 2004: 230). Indeed, this process is largely dependent on the construction of the female colonised body as the ultimate commodity and violable territory – a definition that, in many communities, did not exist before colonisation (Segato, 2016a: 161). As Silvia Federici puts it: "no aboriginal woman was safe from rape or appropriation [...] In the European fantasy, America itself was a reclining naked woman seductively inviting the approaching white stranger" (Federici, 2004: 230). Moreover, since the modern/colonial world universalises a definition of masculinity that is linked to the power over women's bodies, the brutalisation and violation of colonised women by colonised men became a twisted form of *empowerment*. However, while colonised women are particularly vulnerable precisely because they are excluded from the realm of femininity and thus defined as unworthy of protection, their violation itself becomes an ambiguous form of power. In Maldonado-Torres' account:

> Therefore, violence towards the bodies of colonized women can be seen as an affirmation of masculinity that does not carry major consequences. This assertion of violence, however, leaves a hole, as it were, since power over colonized women does not indicate any substantial amount of real power in a system where they do not even properly represent the idea of the feminine.
>
> (Maldonado-Torres, 2016: 17)

These paradoxical relations between sexual violence, power, and the imposition of racialised bodily orders in modernity/coloniality should not lead to the conclusion that the political relevance of sexual violence is any less powerful. If anything, ambiguity only increases the ubiquitousness of sexual violence in the political formation of the flesh. The sexualised rhetoric of conquest, the potency for the destruction of communal bonds, the potential for absolute objectification, and the capacity for symbolic dispossession make sexual violence a fundamental resource in the organisation of the paradigm of the flesh. There is no violence more political, no political order in the modern/colonial system that does not have it at its core.

Conclusion: Violence, Sex, and Politics in the Paradigm of the Flesh

The chapter started with the assertion that sexual violence is always political. As developed throughout the analysis, this statement is important because every order of power, and paradigmatically colonialism (and therefore Western modernity) has instrumentalised violence and sexuality in one way or another. They are not the exterior of politics, but some of its constituting mechanisms for the regulation of the distinctions between valuable and dispensable lives. It is in this regard that sexual violence cannot be thought of as solely oppressive but needs to be understood as a productive practice, which imprints in bodies an entire cosmology of power relations. Thus, this chapter has also argued that this redefinition of sexual violence cannot be separated from a racial and decolonial critique. This is so because the "marks" that accompany the coloniality of gender (Lugones, 2008) cannot be overlooked when addressing the study of sexual violence and/as political violence, and the processes of genderisation, sexualisation and racialisation it inscribes in bodies and practices.

Colonised bodies and subjectivities are indeed shaped by the active interventions of violence. The distinction between the notions of body and flesh, developed by Spillers and more recently reconceptualised by Weheliye (2014), was introduced as the guiding thread of the second part of the chapter because it helps to refocus the question from the centralisation of those bodies that are excluded from the political realm precisely through a systematic and daily exercise of violence. This distinction accompanies and enriches some key dichotomies (political/non-political; human/non-human; gendered/non-gendered) as they organise the modern/colonial order of bodies and power. Attending to this enfleshed surplus of modernity, it is possible to identify how modernity/coloniality imposes a "defence apparatus" that segregates the possibility of violence and self-defence (Dorlin, 2018). What these differential formations impose is the definition of two symbolic realms characterised by the constitution of some bodies as literally 'armed' and entitled to violence, and some other bodies understood as 'open wounds', almost defined through the possibility of violence. Hence, sexual violence itself is the product of the construction of certain positions towards violence and sexuality, but it actively creates these positions: victims and perpetrators are not possibilities that are already given in sexualised bodies, but a relation of bodies and practices that come to be defined, among other things, by the very exercise of violence. And this differential positionality with respect to the possibility of violence sits at the very core of politics.

References

Alcoff, L. M. (2018). *Rape and Resistance. Understanding the Complexities of Sexual Violation*. Cambridge, UK: Polity Press.

Agamben, G. (1998). *Homo Sacer: Sovereign Power and Bare Life*. Stanford: Stanford University Press.

Brown, W. (1995) *States of Injury: Power and Freedom in Late Modernity*. Princeton, NJ: Princeton University Press.

Césaire, A. (2000). *Discourse on Colonialism*. New York: Monthly Review Press.

Crenshaw, K. W. (2017). *On Intersectionality: Essential Writings*. New York: The New Press.

Davis, A. (2011 [1981]). *Women, Race & Class*. New York: Vintage.

Dorlin, E. (2018). *Defenderse: Una filosofía de la violencia*. Trans. By M. Martínez. Buenos Aires: Hekht.

Federici, S. (2004). *Caliban and the Witch: Women, the Body and Primitive Accumulation*. Brooklyn: Automedia.

Foucault, M. (1990). *The History of Sexuality, An Introduction*, vol. 1. Trans. by Robert Hurley. New York: Vintage.

Gautier, A. (2005). Mujeres y colonialismo. In M. Ferro (Ed.), *El libro negro del colonialismo. Siglos XVI al XXI: Del exterminio al arrepentimiento* (pp. 677–723). Madrid: La esfera de los libros.

Hartman, S. V. (1997). *Scenes of Subjection. Terror, Slavery, and Self-Making in Nineteenth-Century America*. New York and Oxford: Oxford University Press.

Khader, S. J. (2018). *Decolonizing Universalism: A Transnational Feminist Ethic*. New York and Oxford: Oxford University Press.

Lugones, M. (2008). The Coloniality of Gender. *Worlds & Knowledge Otherwise*, *2*(spring), 1–17.
Lugones, M. (2010). Toward a Decolonial Feminism. *Hypatia*, *25*(4), 742–759.
Maldonado-Torres, N. (2007). On the Coloniality of Being. *Cultural Studies*, *21*(2–3), 240–270.
Maldonado-Torres, N. (2008). *Against war. Views from the Underside of Modernity*. London and Durham: Duke University Press.
Maldonado-Torres, N. (2016). Outline of Ten Theses on Coloniality and Decoloniality. [Online]. Available at: https://fondation-frantzfanon.com/outline-of-ten-theses-on-coloniality-and-decoloniality/
Oyěwùmí, O. (1997). *The Invention of Women: Making an African Sense of Western Gender Discourses*. Minneapolis: University of Minnesota Press.
Pateman, C. (1988). *The Sexual Contract*. Cambridge, UK: Polity Press.
Pateman, C., & Mills, C. W. (2007). *Contract and Domination*. Cambridge, UK: Polity Press.
Quijano, A. (2000). Colonialidad del poder y clasificación social. *Journal of World-System Research*, *XI*(2), 342–386.
Segato, R. L. (2016a). *La guerra contra las mujeres*. Madrid: Traficantes de Sueños.
Segato, R. L. (2016b). Patriarchy from Margin to Center: Discipline, Territoriality, and Cruelty in the Apocalyptic Phase of Capital. *South Atlantic Quarterly*, *115*(3), 615–624.
Spillers, H. J. (1987). Mama's Baby, Papa's Maybe: An American Grammar Book. *Diacritics*, *17*(2), 65–81.
Weheliye, A. G. (2008). Pornotropes. *Journal of Visual Culture*, *7*(1), 65–81.
Weheliye, A. G. (2014). *Habeas Viscus: Racializing Assemblages, Biopolitics, and Black Feminist Theories of the Human*. London and Durham: Duke University Press.
Wells, I. B. (2014 [1892]). *Southern Horrors: Lynch Law in All Its Phases*. Auckland: The Floating Press.

Part III

Subverting Coloniality

Decolonising the Language of Resistance

6 The Politics of Language in Anti-authoritarian Political Practice: The Southern Mediterranean Case

*Laura Galián**

Introduction

When concepts travel between disciplines, cultures and time, their semantic and socio-linguistic features go with them. Border-crossing makes concepts dynamic and changeable as they travel back, forth, and through, enabling transnational and *glocal* dialogue. Theories travel and are constantly being translated, appropriated, contested, and drafted. In his famous "Notes on Travel and Theory" (1989), James Clifford points out and discusses how travel and theory go hand in hand. Clifford questions the centrality of the concept of 'home' when travelling and confronts it with the idea of periphery and diaspora. These positionalities have an effect on 'theory' which has long been associated with Western discursive spaces. Theory is no longer naturally at home as a powerful site of knowledge when it travels to non-western contexts. This privileged place of 'theory' has largely been contested by other locations and trajectories of knowledge. In this process of displacement and travelling, translation has helped to shift the space of contention of theory while it is being resisted, contested, located, and displaced.

By studying the new linguistic politics and discursive practices in anti-authoritarian movements in the Southern Mediterranean, I want to problematise the univocity of concepts, their history, and their uses (Koselleck, 2012) in order to illuminate new paths in the study of the decolonisation of political concepts. The goal of this chapter is to analyse how different anarchist and anti-authoritarian[1] political activity in the Southern Mediterranean has expanded and reformulated the theory and practice of anti-authoritarian politics through language and translation. All of these countries have different histories and experiences with language, language politics, and decolonisation, but all of them share a search for autonomy and decolonisation through linguistic practices that ultimately break the logic of the colonial authority: the downgrading of colonial languages such as English and French, the introduction of local and marginal languages in their national contexts, subversive and radical translations of Western terminology, etc. All of this reflects the agenda behind

* This article has been funded by the Spanish research proyect: "Emancipatory Concepts in the Mediterranean - Memory, Translation and Transit in their Diachrony" (PID2022-142438NA-100) Agencia Estatal de Investigación

DOI: 10.4324/9781003293460-10

the language and translation ideology of anti-authoritarian movements through texts, messages, and social media presence. Ultimately, this chapter intends to answer the following questions: to what extent do linguistic practices have a direct relationship with the decolonisation of the politics of anarchism? How is this process politically and socio-linguistically articulated in the construction and decolonisation of political concepts?

In this chapter, I study the extent to which anti-authoritarian and anarchist movements and collectives have transformed the discourses and politics of language through their communicative and linguistic practices in the wake of the revolutions of 2011–2019, in various contexts across the Southern Mediterranean. My intention is to understand how the production and circulation of linguistic practices by anarchists and anti-authoritarian collectives have contributed to the decolonisation of political concepts. In order to study these questions, this chapter is divided into two sections. The first part analyses the genealogy of translation, the incorporation of new terminology into the groups' own linguistic and political practices, and the conditions of production and circulation of anarchist theory in Arabic following the 2011 revolutions. The second part examines the politics of language of several anarchist and anti-authoritarian collectives in Lebanon and Morocco, two of the most understudied contexts in the rise of anti-authoritarianism in the Southern Mediterranean.

Translation as a Political Practice of Anarchism

The cultural turn in translation studies in recent decades has brought the translator to the forefront as a creative agent (Pernau & Sachsenmaier, 2016: 15). In translation, ideology plays a significant role in the transmission of knowledge, and for that reason, the act of translating becomes a political act: translators select what to add, what to leave out, which words to choose, and how to replace them with others. Therefore, translation, as a repertoire of political action, reveals the history and the socio-political context surrounding the translator and gives them an entity of their own as a political actor.

As a political practice necessary for anarchist commitments in many parts of the world during the end of the 19th century and beginning of the 20th but also nowadays, translation has helped to decentre and situate (temporally and geographically) the knowledge production of anti-authoritarian practices and thought in the Southern Mediterranean. Translation of anarchist theory and of anarchist terminology is and has been a political practice in itself in the anarchist circles globally, one that is especially relevant in the Southern Mediterranean contexts since most of the literature produced is not originally in Arabic. Going back to James Clifford's notes on theory and travel, when anarchism is translated into Arabic, theory is no longer at home, and 'anarchism' – as a Euro-centred concept and political ideology – is contested and redefined, displaced and negotiated in the Arabic speaking contexts.

My point here is that the process of writing about anarchism in the Southern Mediterranean over the past decade, in particular since the 2011 revolutions, has implied a politicised "translingual practice" (Liu, 1995) where conventions are broken down and conceptual change is actualised. Lydia H. Liu proposes the idea of "translingual practice" to "raise the possibility of rethinking cross-cultural interpretation and forms of linguistic mediation between East and West" (1995: xv). This cross-cultural interpretation in the translation and mediation of political concepts and, specifically, that of anarchism, is at the heart of this chapter. "Translingual practice" is more than mere translation and mediation, its study "examines the process by which new words, meanings, discourses, and modes of representation arise, circulate and acquire legitimacy within the host language due to, or in spite of, the latter's collision with the guest language" (Liu, 1995: 26). According to Liu, translation occurs in this interval where negotiation and contestation take place, and results in languages' transformation. Following Koselleck (2012), once a new concept is created and integrated into a language through translation, there is a change in the experiences gathered by the concept and the meaning of the word itself is updated, altering both the original term and the translated one.

The Arabic language has a long history of translation as contentious politics. *Tarjama* ('translation' in Arabic) involves interpretation and transformation and has played a critical role in intellectual and political projects since the *Nahda* – the Arab 'renaissance' at the end of the 19th century. *Tarjama* is a loanword from Aramaic and Sumerian. It currently means to translate from one language to another. The *Nahda*'s relationship with translation is one of a complex intertwined set of edges that interplay in the colonial contexts of the Southern Mediterranean. In order to understand the relationship between translation and colonial power relations during the *Nahda* period, I use the concept of the "prophetic" following Maya I. Kesrouany. For Kesrouany (2019), the "prophetic" emerges in the translations of the early 20th century, firstly, because the translators saw themselves as prophets of change; secondly, because they translated biographies of the prophet; and thirdly, because the prophetic became a narrative position in their theories of translation. For the author, the "prophetic" undoes the relationship between the translation's origin and destination as a teleological narrative:

> The prophetic, moreover, is *not* a secular project that replaces religion with literature – it does not *pretend* to rid us of the divine. Rather, the prophetic emerges in the space between sacred and secular, original and translation being neither here nor there.
>
> (Kesrouany, 2019: 3)

It is exactly this "being neither here nor there", that is, moving beyond the idea of the starting point and the destination in the task of translating, that I would like to problematise and interrogate in this chapter. 'Translating' or '*yutarjim*', that is, carrying across the layers of meanings and signifiers of anti-authoritarian politics in the Southern Mediterranean, is being done in a neoliberal world whose transfigurations and homogenisation of the political, the social, and the cultural render translation a neglected and secondary activity that is invisible and underappreciated (Hawas, 2012: 278).

However, it also helps to break down and subvert the colonial and neoliberal logic. The politics of translation in anti-authoritarian practices during the 2011–2012 upheavals can be framed in what Mehrez (2012) calls "Revolutionary translation", that is, a collective and non-individual endeavour that serves the purpose of a radical change in the structures of the State and the society. For the members of the Libertarian Socialist Movement in Egypt (LSM), the first anarchist collective established in contemporary Egypt, translation has been part of their understanding of contentious politics and anti-authoritarian political practice. Founded a few months after the 25 January revolution by activists belonging to different generations in different cities throughout Egypt – with two main sections, one in Cairo and one in Alexandria,[2] and with an anarcho-syndicalist and organisational approach to anarchism – the members of the LSM have used translation of (mostly European) anarchist works into Arabic as part of their political practice, organising reading groups to enhance the group's theoretical knowledge. This space has helped to create, share, and compare experiences with what they were living at the time of the social uprisings in Egypt. The need to translate anarchist theory into Arabic was then especially crucial due to the lack of literature on anarchism in Arabic.

In that sense, the blog *al-Anarkiyya bil-ʿArabiyya* (Anarchism in Arabic) has also helped to decentralise and collectivise the task of translating anarchism. Created in 2011 with the emergence of the Arab revolutions, translators and activists from different parts of the Southern Mediterranean – mainly in Egypt and linked to the LSM – have used it to actively engage in the theory and practice of anarchism. Arabic, the normative and standard language variety, is the target language of the blog, which has become an archive that gathers translations of Western and non-Western authors such as Georges Fontenis, Kropotkin, and Graeber, but also the writings of anarchist theorists from the South such as Yasir Abdallah, Mazen Kam al-Maz, Mohamed Jean Veneusse, and Samih Said Abud. The blog has been a reference for the politisation of the younger generations of activists in the Southern Mediterranean that needed a platform of information in Arabic in order to put a name to what they were living and experiencing during the revolution, while also helping to decolonise anarchist political practice through translation.

In fact, when translating the term 'anarchism' into Arabic as 'anarkiyya' (anarchism), 'la-sultawiyya' (anti-authoritarianism) or 'fawdawiyya' (based on the root for 'chaos', it could be translated as 'chaotism'), anti-authoritarian collectives and translators, while embedding their political language within a global imaginary, simultaneously resist the essentialising and homogenising impulse of the neoliberal and colonial politics of language. They do so by, on the one hand, 'domesticating' the translation and using words and concepts of pre-existing narratives and legacies that break through the logic of origin and destination, and, on the other hand, by incorporating new experiences in its meaning into "new translations (ones of potentialities, ruptures and dissonances) that might privilege process over inclusion in a preexisting narrative" (Hawas, 2012: 278).

Anarchist Knowledge Production in a Postcolonial Context

In a postcolonial context, translation and 'traveling theories' entail their own specific political challenges and cultural anxieties. For postcolonial thinker Homi Bhabha (1994), translation occurs at the etymological level when a concept is moved from one place to another. For Bhabha, when this movement takes place in a postcolonial context, the language of the coloniser is subverted and blurred, and the colonial representation is put into question (1994: 33).

It is in this process of movement that 'spaces in between' are created, in which a new language is articulated "through creative performances and cultural and political practices with new semantics, aesthetics and poetics that have created sustained solidarities" (Mehrez, 2012: 14). As noted by Samih Said Abud, one of the most prolific authors and translators of anarchist theory into Arabic and one of the founders of the Libertarian Socialist Movement in Egypt, transnational relations with anarchists all over the world were critical for the re-emergence of anarchism in Egypt in 2011. But, given the centrality of English in these exchanges, educational and cultural capital were also essential:

> I believe that transnational relations with other anarchists have had a great effect on the emergence of anarchism in Egypt and in the Arab world. I know many Arabs and Egyptians in anarchism since then, especially those who have a great knowledge of English, given the scarcity of publications.
>
> (Interview with Samih Said, February 2017)

Therefore, the translation of anarchism in the Southern Mediterranean is not just a tool for democratisation – given that it is understood as a collective endeavour – but also a tool for decolonisation. For Yasir Abdallah, an Egyptian post-anarchist and translator: "Translation helps to liberate anarchism from its European centrality and the theory that emerges from its core. However, the translation must be radical and critical" (quoted in Galián, 2020: 92). In the case of Egypt, many people have learned about anarchism in other languages, such as English, and adopted its terms as an intellectual, cultural, and revolutionary capital. For that reason, argues Yasir, it is important to understand and speak of anarchism in the vernacular language.

I am interested here in theorising the political realities specific to translation, conceptual shift, and the language politics of anti-authoritarian collectives, groups, and individual anarchists in the Southern Mediterranean, as well as the conditions under which the task of translating and the politics of language are presented and negotiated. If we look at the context surrounding the political economy of anarchist translations in the Southern Mediterranean, translation occurs as a situated cultural event that underlies the continuum between the translator and the context of knowledge production. In the case of the Southern Mediterranean, there has been an increase in the publication of translations into Arabic of anarchist theory, as well as the production of anarchist theory directly in Arabic. Ahmed Zaki's 2012 *al-Anarkiyya: al-Madrasa al-thawriyya allati naʿrifha* (Anarchism: The Revolutionary School that We Know)[3] was one of the many publications of anarchist literature in the region since the outbreak of the Revolutions. In this first book, distributed online as a PDF, he reviewed the theory and history of anarchism through the works of the classical authors of anarchism such as Mikhail Bakunin, Nestor Makhno, David Graeber, and others. He also reviewed the most important experiences of anarchism: the Paris commune, the Spanish Civil War, the New Anarchism(s), and the theoretical differences with Marxism. This is how he explained his project to me:

> [...] in 2007, I decided to produce a work in the traditional format, on paper (a book), to present a summary of the anarchist waves at the end of the nineteenth and beginning of the twentieth century from a contemporary point of view and an analytical comparison between the ideological bases of Marxism and anarchism.[4]
>
> (Interview with Ahmed Zaki, 2013)

What is interesting to note is that the book could not be published until January 2011, not only because of the difficulties for emerging writers to break into the Egyptian publishing market, but also due to the topic at hand. However, once it came out in 2011, the book was rapidly and widely distributed during the weeks of the Revolution through the capital's downtown coffee shops and the areas surrounding Tahrir Square, along with copies of translations of Daniel Guérin's *Anarchism: From Theory to Practice*, translated in 1981 by the Lebanese George Saʿad, linked to the Lebanese anarchist group al-Badil al-Taharruri (Libertarian Alternative) affiliated with the French Alternative Libertaire.

Zaki's book was not the first one to be published on anarchism in Arabic, nor the first translation of European anarchist works; however, it was the first one linking the theory with its practice in the region, and the first to be carried out by an activist-theorist and translator. Samih Said Abud, another anarchist from Zaki's generation, describes his encounter with anarchism, that occurred during his formative years, while he was still a member of Marxist organisations in the following way: "All the information I had on anarchism was through the Marxists, who were against anarchism. However, I liked the ideas themselves. There I began to be influenced by the ideas of Rosa Luxemburg and her critique of Leninism" (interview with Samih Said Aboud, January 2017).

Therefore, Marxist organisations' translations of anarchist thought, although funded by the Soviet Union and with a clear intention of opposing anarchism, also helped to create and spread a political language in Arabic in the 1960s and 1970s, during Nasser's years in power. 2011 marked a clear turning point in the proliferation of original and translated anarchist knowledge in Egypt. Activists and non-activists alike suddenly engaged in the process of translating and producing anarchist thought in Arabic, as there was a market that was primed to receive this new literature with open arms. Each of these translations used a different term for 'anarchism', the most common ones in Arabic being, as we have seen, 'anarkiyya' (a transliteration of the Greek word); 'la-sultawiyya' (the combination of the word 'la' (no) and 'sultawiyya' (authoritarianism), therefore, 'no to authoritarianism' or 'anti-authoritarianism'); 'fawdawiyya' (derived from the word 'fawda' (chaos), therefore the reception and conceptualisation of anarchism as chaos); and 'taharruriyya' (libertarianism).[5] While most activists use the terms 'anarkiyya' – with a more Eurocentric perspective – and 'la-sultawiyya' – more focused on the political practice and prefigurative politics than on ideology – academics and translators not directly related to anarchism tend to use the term 'taharruriyya' (closely related to the word 'hurriyya', meaning 'freedom'). 'Fawdawiyya', by contrast, has mainly been used by opponents of anarchism (political and social actors). When looking at the history of the reception of 'anarchism' in Arabic, it is interesting to see that the first translations of anarchism in the liberal newspapers of the 19th century adopted this term, and helped at spreading its semantic meaning as 'chaos' (see Galián, 2020: 95–105; Khuri-Makdisi, 2010: 94–135). However, it has also been taken up by activists who, in an attempt to reconceptualise and change the negative meaning associated with anarchism as chaos, have reappropriated the concept in their writings. The subtitle of Zaki's book stands as an example of this use of 'fawdawiyya' in a way re-signifying its meaning into a powerful political tool for anarchists themselves.

Following Campanella's (2020: 88) understanding of the translation of anarchism, the translations made by anarchists in the Southern Mediterranean break the logic of profit and cultural accumulation, and cannot be reduced to 'propaganda by words' or educational purposes. For Campanella, the idea of profit and accumulation relies on the economy and the market surrounding anarchist publications, that is, their cultural capital. Anarchist translations into Arabic create a literary aesthetic and influence the political processes of the local culture in order to subvert them rather than to create markets and accumulation.

The Language Politics of Anti-authoritarian Practice in Lebanon and Morocco

The very naming and framing of anarchist and anti-authoritarian experiences in the Southern Mediterranean attests to the complexity of significations. Translators and activists are confronted with multiple and undetermined signs and codes. In this section, I intend to analyse the political language of several anarchist and anti-authoritarian collectives in Lebanon and Morocco such as Kafeh!, Tilila, and Hardazat. I chose these two countries for multiple reasons. First, even though both countries were influenced by the revolutions and social uprisings that the region experienced from 2011 onwards, in neither of them did these movements lead to a regime change. In the case of Lebanon, 2011's uprisings sparked what is known as the Intifada of Dignity, which called for political reform and mainly targeted confessionalism in the country. In the case of Morocco, the 20 February Movement, which brought out thousands of people in more than 53 cities to protest against despotism, corruption, and inequality, had a major impact on youth and street politics and led to a constitutional reform in the same year. It confirmed one of the main demands of the movement: the pluralist nature of Moroccan identity and the recognition of Amazigh as an official language.

Second, the two countries have experienced subsequent waves of contentious politics and social uprisings in the years since 2011 that have opened up a window of political opportunity for the proliferation of anti-authoritarian politics and anarchist political philosophy. In both cases, people have revolted against heavily entrenched, corrupt, non-democratic, and authoritarian regimes that have created unbearable living conditions with an outsize impact on the youth. These revolts have fostered the decentralisation, horizontality, and direct action of the previous uprisings. In the case of Lebanon, 2019 witnessed a massive wave of unprecedented protests. Motivated by the direct repercussions of the economic system and corrupt political practices, the protests spread across the country and had a clear non-sectarian, feminist, working-class, and student-based approach. In the case of Morocco, the 2016–2017 Hirak movement – sparked by the authorities' role in the death of fishmonger Mohsen Fikri, who was crushed in the back of a garbage truck while retrieving his confiscated goods – demanded greater economic and social investment in the Rif region by the Moroccan government, as well as the release of political prisoners and the inclusion of Amazigh history in the official history of Morocco (Schwarz, 2019). Third, in both cases, little attention has been paid to these countries' anti-authoritarian and anarchist experiences, whether before or after the 2011 revolutions,[6] compared to the extensive work that has been done on Egypt, Tunisia, and Syria from both a historical and contemporary perspective (Galián, 2015b, 2020; Gorman, 2005, 2008, 2010; Khuri-Makdisi, 2010; Paonessa, 2017; Woller, 2018).

In this section, I examine the politics of language and linguistic practices among the newly emerged anarchist and anti-authoritarian collectives in Lebanon and Morocco, two countries where the multiglossia of the Arabic language is intertwined with the persistent colonial power relations between the languages of the European colonisers and the vernacular languages. How do anti-authoritarian collectives use language? How have the politics of language within anti-authoritarian politics changed since 2011? How do linguistic practices help in decolonising anarchism in the Southern Mediterranean?

Kafeh! is a recently created anarchist group in Lebanon born out of the wave of protests that shook the country in 2019. During the October uprisings that came to an abrupt halt due to the COVID-19 crisis and the port blast on 4 August 2020, Kafeh! by and large, employed Modern Standard Arabic to communicate with its audience on its website[7] and social media platforms,[8] often alongside English. The goal of the group is to achieve a decentralised and anti-authoritarian society based on "liberty, social justice, egalitarianism, and secularism" (Kafeh, 2019). Their presence through digital media, social networks and street politics has been active since then. In fact, the interaction between language and the political is present in the group's very name. In order to explain its meaning, members state that the word 'Kafeh' comes from the expression: "free active anarchist cadres"[9] based on the verb 'kafaha' (to fight against something, to confront something). This play on words in Standard Arabic reflects the importance of the language in shaping this group's identity and translingual practice. It is interesting to observe that, in all instances, the group chooses the term 'anarkiyya' (anarchism) to describe itself, and displays the anarchist circle-A as part of its visual language and semiosphere in its digital and street presence. In contrast to other anarchist groups and individuals that have tended to diverge from the term 'anarkiyya', adopting instead the term 'la-sultawiyya' (non-authoritarianism) and emphasising their political practice rather than its content, Kafeh!, by choosing 'anarkiyya', do not distance itself from anarchism's European background and canon.

Kafeh! exemplifies the translingual practice of the October revolution in multilingual Lebanon, where new social movements and identities have created new forms of linguistic 'conflict', while bringing into view new subjectivities in the political sphere. In Lebanon, as in other parts of the Arabic-speaking world, the variety of Arabic used for writing is almost exclusively Standard Arabic (SA). While Lebanese Arabic (LA) is the most widely spoken Arabic variety and the mother tongue of most Lebanese people, Standard Arabic is the official language of the state. Another layer of the country's multilingual situation is the use of French and English, both of which are related to the recent history of colonisation.[10] However, while French, the traditional colonial and elite language, has lost vitality in Lebanon, English, as the language of education and global communication, is on the rise (Iriarte Díez, 2021: 8). In addition to LA being the most commonly spoken variety, a form called 'Arabizi' – colloquial Arabic written with Latin script – has recently emerged among young people, emphasising the link between Arabic script and SA.[11] This being the case, the intersection between language use and ideology continues to be an important tool and marker in building a political identity and culture, as exemplified by Kafeh! whose members, interestingly, do not include LA in their digital media interactions, despite the many slogans in LA that were in circulation during the October revolution.

The October 2019 uprising in Lebanon was a turning point across the political, social, and linguistic domains. The announcement of a new tax on WhatsApp calls, gasoline, and tobacco brought a significant section of the population out into the streets to condemn the corrupt political system and call for an end to sectarianism. In fact, with the clear intention of overcoming socioeconomic divisions and reclaiming the public space against a common enemy – the economic and political elite – the street movement saw an unprecedented mass demonstration where gender issues came to the forefront – at times even more than class issues. As Ana Iriarte Díez points out in her study on language attitudes in Lebanon during the October revolution, the status of Arabic has noticeably shifted in parts of Lebanese society (2021: 13). Arabic was the language through which Lebanese people made their demands, wrote on banners, walls, and chanted against corruption and injustice. However, it is difficult to establish which Arabic from the spectrum of 'Arabics' was used, since the mix between varieties is a common feature of the language's use. It seems that one of the motives that the speakers had in using Arabic after the October revolution was to make it more inclusive and overcome social and economic differences (Iriarte Díez, 2021: 29). Therefore, language is used as a marker and practice of contentious politics towards social equality in daily and political exchange. However, while Arabic creates a collective identity among 'the people' (*al-sha'ab*), since the revolutionary message can cut across social classes and sects, I contend that Kafeh!'s choice of SA and the devaluation of other varieties of Arabic used in Lebanon responds to two factors: 1. LA is still perceived as the language chosen by Lebanese nationalism, and anarchism, as an internationalist idea, is against the exclusivity and sectarianism this would suggest; 2. SA is the language of this internationalisation, since it also has a reverberation in other Arabic-speaking countries and, at the same time, is not the language of the European colonisers (English or French).

In the case of Morocco, the 2000s marked the transition of Darija – for centuries the spoken language of Morocco – to a written language. Before this time, written Darija on the internet mainly occurred in texts written by intellectuals.[12] However, it was the 20 February Movement, inspired by the revolutions that were taking place in the region in 2011 and 2012, that gave Darija its final push into the language of protest and social movements. It is important to understand that Moroccan Darija was not constitutionalised like Amazigh was in the 2011 Constitution, which was drafted to defuse social and political tensions, nor has it been officially recognised as a component of Moroccan culture (Kettioui, 2020: 13). Furthermore, postcolonial contexts such as Morocco have been dominated by elites whose fluency in French and SA has acted as a gatekeeper to positions of power. In that sense, urban social movements and youth identities have popularised the use of Darija as a subversive and politicised language. According to Kettioui, the development of Darija can be traced as an oral and written language. First as the oral language of poetry, second as the language of urban youth music, and third as the language of the 2011 uprisings:

> The controversy around language is not a linguistic one per se. More broadly, it is an expression of a youth subculture reconstructing its postcolonial subjectivity at the interstices of tradition and modernity. These authors in Darija deploy autoethnography to sketch everyday social, political and economic violence, miseries, ironies and subtleties haunting an increasingly neoliberal and globalized postcolony.
>
> (Kettioui, 2020: 4)

In the following autonomous, decentralised, and anti-authoritarian collectives based in Morocco – Tilila and Hardazat – Darija acts as a third space between the language of the elite and the colonisers. Neither of these groups describes itself as 'anarchist', though they emphasise anti-authoritarianism and decolonisation as basic principles. Tilila describes itself as a group of "young Moroccans who aim to act, debate and give concrete solutions to improve the educational system in Morocco" (Tilila, 2018). Although founded as the Union of Students for the Change of the Educational System (UECSE) in 2012, it changed its name to Tilila in 2017.[13] And despite its primary aim to fight for the improvement of the education system in Morocco, its main goal is now phrased in the following terms: "bhrina mujtama'a mutaharrir, mutaddamin wa qad birasu" (We want a free, inclusive and empowered society) (Tilila, 2018). In fact, their objectives are direct democracy, anti-authoritarianism (la-sultawiyya), liberation (al-taharrur), the struggle against all forms of discrimination (munahida al-tamyiz), and intersectionalism (al-taqatu'ayya). Based in different cities around Morocco – mainly Rabat, Casablanca, and Agadir – the members of Tilila organise cultural, educational, and artistic activities that engage the youth in problems related to their political and environmental surroundings. The language used by the collective on its website (mainly through Facebook) reflects the horizontality of the organisation and its intention to elevate the status of Darija, but also its historical and *glocal* connections. French and English, to a lesser extent, are used in its social media together with written Darija (and Amazigh in a more marginal way) as the main languages of communication. Although most of their publications are disseminated in both English or French and Darija, it is Darija that acts as the language of politics (in their communiqués), culture, and the arts, thus elevating the use of the 'vulgar' to the language of daily-life concerns and political theory: "The voices of Morocco's post-uprising literature in darija have silently, diligently and steadily mainstreamed darija, the language of the 'unwashed masses', into the web's first mode of expression through everyday writing" (Kettioui, 2020: 7).

Tilila is not the only anti-authoritarian group that privileges the use of Darija. In the last few years, there has been an emergence of Temporary Autonomous Zones (TAZ) connected to music. Born as a decolonial music festival against racism and white privilege, a safe space for people of colour, "where the violence they're victim of due to the supremacy of northern culture over countries of the South is not tolerated" (Hardzazat Hardcore Fest, 2019), Hardzazat Festival held its first edition in 2018 in Ouarzazate, located in Southern Morocco. With the goal of creating a safe space for people from the Global South to talk about racism and organise against white privilege, sexism, and homophobia, and to build an "autonomous movement of revolted and revolutionaries" (Hardzazat Hardcore Fest, 2019), Hardzazat Hardcore Festival uses, like Tilila, both Darija and English (and at times also French) in their digital media and linguistic practices. Although, contrary to what one might expect, English is the only language used on their website, written Darija, Arabizi, and English are together the most commonly used languages on their social media platforms. As noted by Kettioui (2020), Darija has emerged as the language of the underground among Moroccan music festivals. Rap, punk, heavy metal, and social media platforms have given Darija new momentum.

The use of Darija exemplifies the subversion of the State power and institutions in the country since SA is the language of the Makhzen (the Moroccan state apparatus), which is legitimised through family and religious ties. For Morocco's anti-authoritarian social movements, the use of Darija is also a reflection of consciousness operating against the hegemony of accepted norms. In fact, the intention of the Hardzazat Festival is to understand and subvert deeply entrenched cultural norms, arguing that these ideas are part of a "fascist ideology" that constructs "deviance" (Hardzazat Hardcore Fest, 2019). Therefore, the use of Darija by these new anti-authoritarian social movements opens up a third space, a linguistic transition from below creating a new language where the 'vulgar' becomes the vehicle of the expression of the youth and the masses, but also constitutes an anti-fascist political practice: "[...] this new Darija strand is revolutionary not in the sense of mobilizing people against the state, but in the sense of carving out a new literacy, literality, sensitivity and counter-public" (Kettioui, 2020: 15).

Conclusion

Although not all movements analysed in this chapter recognise that language and translation constitute spaces of resistance, in their translingual practice they go beyond language as a signifier of nationalist aspirations or religious elitism, but articulate it instead as a narrative of global justice within a local practice. In that sense, they work within the motto of "thinking globally, acting locally" of the New Social Movements that emerged after Seattle 1999. They are *glocal*, that is, they

> [...] require a global reconversion of the processes of socialization and cultural inculcation and of development models, or they require concrete, immediate and local transformations (for example, the closure of a nuclear power plant, the construction of a day-care center or a school, the prohibition of violent advertising on television), demands that, in both cases, go beyond the mere granting of abstract and universal rights.
> (Sousa Santos 2001: 180)

By going beyond the logic of Universalism in their linguistic practices, these movements and collectives break the logic of modernity and decolonise their practices. In the words of Heba Raouf, "they introduce a new a-modern logic" (2004: 41), not by deliberately downgrading English, as the language of the coloniser or the language of the neoliberal order, but by allocating a different order in the space they ascribe to the languages that interplay in their *glocal* position. Contrary to the examples analysed by Mona Baker (2013) of activist translators within the framework of New Social Movements and revolution in Egypt that deliberately undermined the use of English in relation to other local or vernacular languages, the anti-authoritarian or anarchist collectives analysed in this chapter use English or French, together with standard Arabic, vernacular Arabic or other minority languages (such as Amazigh), depending on the audience targeted in their messages since they are all languages that form part of a global community. Likewise, they see themselves as part of a transnational network of resistance and solidarity, and as local agents in this broader struggle for decolonisation. As in the case of 20th-century Argentinian anarchism, the linguistic practice of Southern Mediterranean anarchism(s) "does not associate each language with a national identity, nor does it adopt the polyglotism of the elites, but rather the opposite" (Campanella, 2020: 85). As Marianna di Stefano states, in the multilingualism of anarchists "this set of languages in simultaneous use is related to proletarian internationalism" (cited in Campanella, 2020: 85). In the case at hand, rather than the "proletarian internationalism" of the 19th and 20th centuries, here, the use of multilingualism is related to the global language of revolt against the neoliberal and capitalist order, but also against the local power relations.

In an attempt to avoid the notions of centre and periphery, their language ideology places non-European theories, concepts and languages on equal footing with their European counterparts, and puts works in translation on par with their originals. While the translation of anarchism becomes a mandatory response to the neoliberal order and persisting colonial power relations, Arabic plays a central role in their understanding of anarchism. For that reason, translation practices, that is, what words translators choose – such as that of 'anarkiyya', 'la-sultawiyya', and 'fawdawiyya' as explained along the chapter – have an important effect on how this political philosophy is understood and practised in the South of the Mediterranean. On the other hand, the politics of language as a practice of anarchism – that is, why do movements choose one language over the other in specific contexts, how do they work, and how does it impact the practice of anti-authoritarian politics – does also help at questioning the deeply rooted language of neoliberalism, colonialism and capitalism and therefore ultimately helps at decolonising political concepts. In some contexts, such as Morocco, local non-prestige languages such as Darija and Amazigh have also emerged as part of the decolonisation of political concepts, in a growing context of subversion of SA as the language of the political and religious elites.[14] By doing so, new concepts are being formed, actualised and re-semanticised; not only in Arabic, but also in the understanding of anarchism globally by incorporating new experiences from the Southern Mediterranean into the overall genealogy of the concept.

In this change in the locus of enunciation lies the importance of becoming at the same time a theoretical and practical corpus, that is to say, they are "both in the place of oppositional practice in the public sphere and that of a theoretical struggle in the academy" (Mignolo, 2005). This duality of enunciation from the South, such as the issue of how to refer to 'anarchism' in Arabic or Arabics, is what subverts the hegemonic order and holds the potential to be both weapon and space of subversion, wherein the very act of theorising constitutes a political act. Therefore, the classic theory/practice dichotomy into which anarchism has traditionally been divided has become obsolete in the postcolonial context. If the act of 'enunciating' is a political act – because of the dual aspect of these disciplines – the act of naming, translating, and theorising anarchism is, depending on who does it and from where, also an anarchist practice that ultimately serves to decolonise anarchism.

Notes

1 Although in this chapter I understand anarchism as an etiquette or a self-proclaimed ideology and anti-authoritarianism as a way of doing politics, at times they will appear as interchangeable.
2 For more information on the Libertarian Socialist Movement in Egypt, see Galián (2015a).
3 In the first publication of the book, distributed online in PDF format in 2007, its title was *Dif'an 'an al-fawdawiyya. Al-Madrasa al-thawriyya allati lam na'rifha* (In Defense of Anarchism: The Revolutionary School That We Did Not Know).
4 Translations are my own unless otherwise noted.
5 For a more detailed explanation of the process of constructing the Arabic terminology of anarchism, see Galián (2020: 91–134).
6 In the case of Lebanon, I know of the existence of al-Badil al-Taharruri (Libertarian Alternative), a branch of the French Alternative Libertaire, which operated between 1995 and 2008. It was, as far as I know, the first anarchist group in the Southern Mediterranean in contemporary times. In 2008, the anti-authoritarian collective Radical Beirut emerged as a response to the attacks on Gaza, and was active mainly on social media, but also in various actions on the street. In the case of Morocco, I do not have enough evidence of the existence of anarchists or anti-authoritarian groups before 2011, which does not mean that they did not exist. The French anarchist group Alternative Libertaire, as in the Lebanese case, had a section in Morocco that was affiliated with the 20 February Movement.
7 Although I have visited their website several times, the link currently seems to be broken (retrieved from kafeh.org, 6 November 2021).
8 See the Facebook and Twitter accounts: https://www.facebook.com/kafeh.lebanon/, https://twitter.com/kafehleb?fbclid=IwAR1G_DSkMsDKld-ZO9MRIuMxhUKW4NPP9xFGHFViRy8w8I97G-xNYVoeGV4
9 See https://organisemagazine.org.uk/2019/10/08/kafeh-announcement/
10 French was the educational language enforced during the Mandate in Lebanon (1925–1943) and the 1926 constitution imposed French as an official language and the language of instruction for scientific subjects. English, on the other hand, was first brought by American Evangelical Protestants, in the late 19th and early 20th centuries, who spread the language via academic institutions.
11 However, the use of Arabizi is now not as common as LA.
12 With the advent of pan-Arabism and the tensions that arose with Moroccan independence in 1956, Darija has been considered a vulgar language as compared to Quranic Arabic.
13 Tilila started as the Union des étudiants pour le changement du system éducatif (Union of students for the change of the education system) in 2013. This collective, which arose spontaneously, had the aim of "constituting a unified front against the disastrous policies applied in Morocco since independence until today in the field of education" (Vogel, 2013).
14 Noting that, however, many intellectuals and activists have also maintained a commitment to SA.

References

Baker, M. (2013). Translation as an Alternative Space for Political Action. *Social Movement Studies: Journal of Social, Cultural and Political Protest*, *12*(1), 23–27.

Bhabha, H. K. (1994). *The Location of Culture*. London and New York: Routledge.

Campanella, L. (2020). La traducción como práctica política: Les 21 Jours d'un Neurasthénique de Octave Mirbeau En el periódico anarquista Nuevo Rumbo. *Mutatis Mutandis. Revista Latinoamericana de traducción*, *14*(1), 68–91.

Clifford, J. (1989). Notes on Travel and Theory. *Cultural Studies*, 5. Retrieved from: https://culturalstudies.ucsc.edu/inscriptions/volume-5/james-clifford/

de Sousa Santos, B. (2001). Los nuevos movimientos sociales. *Revista del Observatorio Social de América Latina/OSAL*, *5*, 177–188.

Díez, A. I. (2021). Language and Revolution: Arabic in Lebanon after October Revolution as a Case Study. *Journal of Arabic and Islamic Studies*, *21*, 5–37.

Galián, L. (2015a). New Modes of Collective Actions: The Reemergence of Anarchism in Egypt. In F. Gerges (Ed.), *Contentious Politics in the Middle East. Popular Resistance and Marginalized Activism beyond the Arab Uprisings* (pp. 351–372). New York: Palgrave MacMillan.

Galián, L. (2015b). Hacia un estudio decolonial del anarquismo. Perspectivas comparadas de Egipto y Túnez. *REIM*, *18*, 44–73.

Galián, L. (2020). *Colonialism, Transnationalism and Anarchism in the South of the Mediterranean*. Granada: Palgrave MacMillan.

Gorman, A. (2005). Anarchists in Education: The Free Popular University in Egypt (1901). *Middle Eastern Studies*, *41*(3), 303–320.

Gorman, A. (2008). Socialisme En Égypte Avant La Première Guerre Mondiale: La Contribution Des Anarchistes. *Cahiers d'histoire. Revue d'histoire critique*, *106*(2008), 47–64.

Gorman, A. (2010). Diverse in Race, Religion and Nationality... But United in Aspirations of Civil Progress: The Anarchist Movement in Egypt, 1860–1940. In S. Hirsch & L. Van der Walt (Eds.), *Anarchism and Syndicalism in the Colonial and Postcolonial World, 1870–1940* (pp. 3–31). Leiden and Boston: Brill.

Hardzazat Hardcore Fest. (2019). Decolonial & Antifa Fest. *Hardzazat Hardcore Fest*. Retrieved from: https://hardzazat.wixsite.com/hardzazat/decolonail-antifa-fest

Hawas, S. (2012). Global Translations and Translating the Global: Discursive Regimes of Revolt. In S. Mehrez (Ed.), *Translating Egypt's Revolution: The Language of Tahrir* (pp. 277–306). Cairo and New York: American University in Cairo Press.

Kafeh! (2019). About Us. *Kafeh!* Retrieved from: https://kafeh.org/about-us/

Kesrouany, M. I. (2019). *Prophetic Translation. The Makin of Modern Egyptian Literature*. Edinburgh: Edinburgh University Press.

Kettioui, A. (2020). Sarcasm and Taboo in the Moroccan Mediascape after the February 20 Movement. *Journal of African Cultural Studies*, *33*(4), 405–423. doi:10.1080/13696815.2019.1701426

Khuri-Makdisi, I. (2010). *The Eastern Mediterranean and the Making of Global Radicalism, 1860–1914*. Los Angeles and London: University of California Press.

Koselleck, R. (2012). *Historia de Conceptos. Estudios sobre semántica y pragmática del lenguaje político y social*. Trans. by L. Fernández Torres. Madrid: Editorial Trotta.

Liu, L. H. (1995). *Translingual Practice. Literature, National Culture, and Translated Modernity - China, 1900–1937*. Palo Alto, CA: Stanford University Press.

Mehrez, S. (Ed.) (2012). *Translating Egypt's Revolution: The Language of Tahrir*. Cairo and New York: American University in Cairo Press.

Mignolo, W. D. (2005). La razón postcolonial: Herencias coloniales y teorías postcoloniales. *AdVersuS: Revista de Semiótica*, 4. Retrieved from: http://www.adversus.org/indice/nro4/articulos/articulo_mingolo.htm

Paonessa, C. (2017). Anarchismo e Colonialismo. Gli Anarchici Italiani in Egitto (1860–1914). *Studi storici: Rivista trimestrale dell'Istituto Gramsci*, *58*(2), 401–428.

Pernau, M., & Sachsenmaier, D. (Eds.) (2016). *Global Conceptual History. A Reader*. London and Oxford: Bloomsbury.

Raouf Ezzat, H. (2004). Beyond Methodological Modernism: Towards a Multicultural Paradigm Shift in the Social Sciences. In A. Helmut, M. Glasius & M. Kaldor (Eds.), *Global Civil Society 2004-2005* (pp. 40–58). London: SAGE.

Schwarz, C. H. (2019). La socialización política transmediterránea: El movimiento Hirak, la diáspora marroquí y Europa como imaginario político. *Quaderns de la Mediterrània*, *28–29*, 280–287.

"Tilila -تيليلا." (2018). *Facebook*. Retrieved from: https://www.facebook.com/collectif.tilila/?ref=page_internal (15 November, 2021).

Vogel, M. (2013). Penser Le Changement de l'éducation Au Maroc: Interview de Nabil Belkabir, Membre de l'UECSE. *Jadaliyya*. Retrieved from: https://www.jadaliyya.com/Details/27808 (15 February, 2022).

Woller, A. (2018). Self-Mediation Practices of Arab Anarchists. In C. Richter, A. Antonakis, & C. Harders (Eds.), *Digital Media and the Politics of Transformation in the Arab World and Asia* (pp. 35–60). Berlin: Springer VS.

7 Decolonising Sovereignty and Reimagining Autonomy

Adivasi Assertions and Interpretations of Law

Astha Saxena and Radhika Chitkara

Prologue: A Story from Forests

Disobedience, struggles, and political assertions have been an integral part of Indian democracy (Guha, 2017). In these historical processes, people across the country have consistently challenged the established order to make these assertions. One such process has, yet again, emerged from the forests of India under the name of "Pathalgarhi movement". The movement's form and content are unique in that the communities that lead it claim rights over their lands by making the Constitution their own and providing interpretations to it that have seldom been heard before. Assertions made by these forest-dwelling communities may not be accurate in law or polished in language, but they symbolise something deeper: an idea of autonomy which forms the basis of our democratic process and of our identity as a nation.

In this chapter, we unpack the idea of autonomy in the Indian context and examine its relationship with state sovereignty – the differences in their forms of power and manner of assertions. There are at least 40 million hectares of forest land eligible for Community Forest Rights under the 2006 Forest Rights Act (FRA) (RRI et al., 2015). Meaning, that there is an expansive area across the country that can legally fall under the autonomy of Adivasis, forest-dwelling communities, and indigenous communities.[1] The FRA is a paradigm-shifting legislation of postcolonial India that seeks to undo the historical injustices of the past, particularly of the colonial period when forest communities were alienated from their lands (Sarin & Springate-Baginsk, 2010). In this context, we attempt to decolonise sovereignty in order to break the distinction created between it and autonomy. Adivasi and forest-dwelling communities of India have had a historical relationship with forests and land, their identity and existence are tied to it. For them, material alienation, such as the loss of forest, is equivalent to a complete loss of existence. In the current emerging trends of neoliberalism where land alienation has gained new meanings and processes, Pathalgarhi occurs as another assertion of the communities to retain their identity. Therefore, legal theory needs to go beyond simply examining the *legality* or *illegality* of claims coming from political struggles. Instead, it needs to bore

DOI: 10.4324/9781003293460-11

into the conceptual underpinnings of autonomy and sovereignty to reveal existing limitations of thought. We begin here with one such story from forests.

In the heart of the Jharkhand homeland, visitors are greeted with tall green stone slabs by the Gram Sabhas[2] of the village, declaring the law applicable in the area by virtue of the Fifth Schedule of the Indian Constitution.[3] These 'pathals', or stone slabs, declare:

1 That, by virtue of Article 13(3) of the Indian Constitution, traditional and customary laws of Adivasis have the force of law in the area;
2 That, under Article 19(5) of the Indian Constitution, the autonomous Gram Sabhas of the village hereby prohibit the entry, residence, settlement and free movement of non-Adivasis in the village/ area;
3 That, vide Article 19(6) of the Indian Constitution, the autonomous Gram Sabhas of the village further prohibit the conduct of business and trade by outsiders in the village/ area;
4 That, by virtue of Article 244(1) and the Fifth Schedule, laws enacted by the Parliament and State Legislatures do not have the force of law in the village or area.[4]

Over the course of 2017–2019, more than a hundred villages in areas like Ranchi, Saraikela Kharswan, and Latehar in the state of Jharkhand – later extending to Chhattisgarh, Odisha, and Madhya Pradesh in different texts and forms – erected these stone slabs along the customary boundaries of their villages as part of the Pathalgarhi movement. The same Munda[5] homelands had historically been sites of militant assertion of Adivasi rights over *jal*, *jangal*, and *zameen*.[6] These communities believe that jal (water), jangal (forest), and zameen (land) are three connected elements in nature and their existence and identity are also connected with these three elements combined. In the language of these communities, their claim is not to land or forest alone, but to jal, jangal, and zameen as one, which without the others, is no right at all (EPW Engage 2019). 'Pathalgarhi', literally translating to 'erection of stones', has historically been a cultural practice of marking sacred spots over time symbolising the Adivasi homeland. These *pathals* gradually took the form of political assertion over customary lands in post-Independence India, coinciding with the rapacious loss of lands since the 1980s to industrialisation and major development projects.[7] But, when a major conference of foreign investors called 'Momentum Jharkhand' was held in 2017 to invite domestic and international corporations to invest in Jharkhand on the promise of easy and plenty availability of land, Pathalgarhi was reborn as a movement of resistance against state and economic forces. Already, more than 8500km^2 of rainforests had been transferred by the state government into 'land banks' – without informing, consulting or seeking the consent of the Gram Sabhas – in order to facilitate business for these investors. Drawing from the powers and authorities of self-governance under the Indian Constitution, Gram Sabhas rejected state authority and initiated a return to self-governance. The state response to this movement was spearheaded by the criminal justice system leading to the arrests of at least 10,000 people (Sharma, 2019; CDRO-WSS, 2018).

Even as Adivasi communities' exercise of their political rights stands curtailed under the pressure of paramilitary deployment and coercive criminal processes, the genie is out of the bottle and has unleashed a contest, simplistically put, between 'formal' and 'popular' constitutional interpretations. In a dialectical tug of war between constituent autonomous units and the state, who decides?

Contextualising the Debate

To a world that dwells in definitive ideologies and is crowded with preconceived colonial notions, this story would have had an obvious response. Assertions that oust the authority of the Parliament and the state legislature would lie outside the realm of law – that they would be, at best, unjustified political aspirations of a people who seek to reject democratic institutions and oppose peace and the common good of the nation. A considerably reasonable remark might concede the legitimacy of demands for land rights but would tag this mode of resistance as illegitimate, and even unconstitutional. Contextualising this story, however, cannot possibly be determined by a political-legal discourse that presumes certain standards of legitimacy and legality. In the postcolonial context, it may be apparent to label the use of state violence as legitimate and the peoples' interpretation of the law as illegitimate, but that would also be substantially perfunctory. For, a process of contextualisation demands intellect to be free and disposition to be empathetic. Merely because the interpretations given to the Constitution by the Adivasi communities of Jharkhand in the Pathalgarhi movement are not formally made, they cannot be rendered illegal or incorrect. However, because our perceptions are still stuck in colonial compartments of 'civilised' worlds and 'sovereign states', we often fail to see the justness in these claims and are instead entangled in questions of legality.

The principle of sovereignty has, along with framing national identities, precepted our spectrum of political knowledge, imagination and definite structures of authority (Shaw, 2008). In its paradigmatic formulation, the principle is also traversed by the history of coloniality: much like in the case of India, for a nation to *become* sovereign was equivalent to attaining freedom from colonial rule and establishing its own government by taking guidance from Western principles and models. Sovereignty establishes a common political order that commands dominance as the arbiter of the common good and the dominant legal system of the state remains deeply entrenched in the notion of a singular ultimate rule-making authority over a territory. This necessitates the extinguishment, and to a limited extent accommodation, of certain *others* (Shaw, 2008) – of authorities and visions deemed not sufficiently civilised or too backward to accept the common good of the modern state. Although political and legal theorists have devoted centuries to ascertaining a rational basis for this sovereignty,[8] its source of authority is still contested. Even so, we find ourselves thoroughly believing in the worldviews and structures established by it, while forgetting our own individual and community sovereignties (Shaw, 2008). We tend to forgo its colonial associations and ascertain the sovereignty of the nation; as if, it were possible for this call for common good to be disjunct from its history. The Indian freedom struggle represented a structural split between the elite and the subaltern domains of politics and therefore, the consequent sovereignty that has been established needs to be investigated (Guha, 1998). Our attempt in this chapter would be to reimagine these conceptualisations in order to make space for decolonial articulations of Law and of the political relationships between states and peoples mediated through the operation of sovereignty and Law.

As law researchers by training, we are obligated to confront the presumptions of this discourse. Our primary concern here is the internal dimension of sovereignty in its *legal* sense, in the creation, legitimation, and exercise of state authority over rule-making and enforcement. Our question is: is sovereignty an idea of *singularity*, and possibly hegemony, or is there a fundamental element of *plurality* in the concept, and thereby a space for negotiation and co-constitution? While the emergence of the nation-state and sovereignty may be contemporaneous, existing critiques do not necessarily interrogate the tensions produced by the formal creation of sovereignty as it inscribes its territory over pre-existing people and negates their rule-making authority over land and resources. This produces tensions for democracy between the ideas of sovereignty as the singularity of political authority within a territory and as an expression of the inherent right of peoples to self-determination within a territory. There are many different ways of describing 'sovereignty', but we focus on its relationship with other modes of power and authority within the same territory. We find this exercise important despite the proclamations of the death or irrelevance of sovereignty (Roth & Mullins, 2010).

Keeping these questions as our substructure, this chapter is dedicated to investigating the terms of postcolonial sovereignty, the place it acquires in theory, and how it engages with indigeneity. It seeks to contextualise the story from forests in a manner that neither rejects nor paternalises, but lends an ear in fairness. And if our aim is to have any decent engagement even remotely with Adivasi experiences like the one narrated in the beginning, then settled principles will have to be unsettled and interpretations would need to be reimagined. We will have to perceive the stone slabs differently. In other words, we will need to confront the predominant idea of singularity and to actively decolonise the conceptual boundaries of sovereignty, to unearth the colonial connotations and linguistic limitations that surround it. The writings on pathals are a language of resistance that cannot be perceived as only *illegal*. They need to be understood for what they are: assertions of autonomy by people who were historically promised such power.

This chapter is divided in four parts. We began by narrating the story of resistance coming from forests and the ways in which it has been perceived. It allowed us to introduce the theoretical limitations surrounding such languages of resistance and what is required to contextualise them. The second section is an inquiry into the difficulties of a political discourse that centralises sovereign power and concedes all others to be an alternate arrangement, especially when the question of Adivasi identity is opened. This part describes the inherent difficulties in practising binaries and dwelling in dichotomies, characteristics of colonial thinking, and argues that our world is instead characterised by pluralism. The third section looks at the ways in which the concept of sovereignty can be decolonised by investigating the possibility of forging horizontal relationships between sovereignty and autonomy. This relationship is often lost in the unitary colonial conception of sovereignty. We argue that given the nature of these powers, they are bound to be construed in a horizontal and thereby in an equal relationship, instead of a vertical one popularised and dominated by the state. The fourth section takes a dive into the Indian constitutional arrangement to see the various forms of sovereignties and autonomies that have been established to sustain the Indian state. In this concluding section, we unfold the possibility of lending an ear in fairness to Adivasi and forest-dwelling communities as they leap to offer creative interpretations of the law that governs us all. The chapter ends by arguing that the starting point for this process of decolonisation already exists in the Indian Constitution and that, what is needed is for the legal community to have an open heart and an open mind.

The Difficulties of Discourse: Asserting Dominance and Precluding Alternatives

One of the many difficulties in writing about Adivasi concerns is that one is unsure of where or how to begin. A simple enunciation of problems associated with the Adivasi discourse can hardly be helpful. Conversely, any implication that the discourse is largely unproblematic and that most prevailing difficulties are operational would be offering a false premise.[9] One also cannot simply assert that Adivasis need to move with time and adapt to the ways of a modern developing world – for terms of *movement* or *modernity* or even *time* are not universal constants and are profoundly entangled with coloniality (Dussel, 1995; Quijano, 2007; Mignolo, 2007; Roth & Mullins, 2010). As significant as challenges in implementing existing laws may be, the norms, institutions, and processes crafted for the distribution of land and resources themselves suffer from foundational difficulties. This is the central contention we advance in this chapter.

While constantly creating and maintaining a world of right and wrong, our imaginations seem to conclude within the rules and principles of such dominant notions of justice. We create our own versions of truth or of justice which are manifested in the building of formal institutions within the nation-state format; one that is created through dominant versions of coloniality and has crystalised in our minds (Wright, 2016). All our concepts, terms of engagement and even boundaries of knowledge are inseparable from our narratives (or experiences) that give them meaning. Although significant, they are, however, but a small part of the normative universe that ought to claim our attention

(Cover, 1983). Communities, and Adivasi in particular, consistently create and manage these universes, producing legalities of their own and meanings that are very different. They have inhabited parts of this world and managed forests and resources for centuries, thereby creating their own historicity of ideas and social order. Their realities of concepts are as real as any other one. These conceptualisations are, therefore, nothing but a variant representation of justice that are created and maintained as a part of the same world as anyone else (Davies, 2017). The idea that the paradigmatic conceptualisation of politics, sociology, economics, and law is only one of the many requires us to address our limitations of language, of experience, and of our perceptions of right and wrong, along with our identities that are created in the process (Davies, 2017). Adivasi assertions of autonomy are a powerful entry point to these questions, for they open for us a space to think outside of the dominant. The existing affiliation of legitimacy with certain specific forms of governance and sovereignty is rather baffling; they are, after all, only one solution to the larger problem of social order. However, the political thinking sustained by modernity/coloniality seems to require illegitimating the other to obtain our sense of orderliness.

So, it becomes pertinent to cut across our layers of perception and understand how dichotomies of right and wrong have been formed. How is it that the anchor of dominant theories assumes Adivasi systems to be the *others* with an *alternate* means of living? How is it that perceptions have been consistently formed around the binaries of *right–wrong*, *valid–void*, *legitimate–illegitimate*, and *legal–illegal* and that concepts are constructed upon a limited understanding of these binaries? And how has the postcolonial perception affected these binaries? Ideas and processes are either considered correct or incorrect based upon their source of authority, making Adivasi claims usually fall in the latter category. Consequently, all political and legal systems are built around these 'correct' ideas, necessarily ousting any variant view on the matter. The concepts of nation-state, sovereignty, government, governance, self-determination, autonomy, democracy, federalism, eminent domain, ownership, development, and welfare are all constructed with largely one legal notion of the world. These conceptualisations reflect coloniality's aspiration and imposition of unicity, as the politics of purity working against plurality (Maldonado-Torres, 2018; Monahan, 2011). Any idiosyncrasy of a community, multiplicity of meanings, or overlap of legalities do not form a part of this description; they are either illegitimate or undeserving of any attention. How then, does one articulate Adivasi concerns in a scenario where they have already been titled *incorrect* based upon their point of origin or a jurisprudential problem of methodology?

In this process of articulation, the leading difficulty arises in the treatment of Adivasi discourse as an exception occurring in a world of norms and order. Therefore, Adivasis, their ways of living, and their perception of governance and of development are considered relevant in contexts that are only talking about their rights. However, the questions and concerns raised by the Adivasi discourse cut across the very foundations of our paradigms of understanding. Article 13 of the Constitution of India is often interpreted for colonial laws to stand the test of the fundamental rights. But when this provision is written down on a stone slab by an Adivasi community as we saw in the beginning, it achieves a different meaning; one that reconstructs autonomy within the constitutional framework. The community thus participates in giving meaning to the Constitution. When Adivasi concerns are spoken in a context of speciality or exceptionality, any kind of normalcy is denied. Every aspect of Adivasi experiences is then discussed only in spaces meant exclusively for them, as a matter of their rights that have to be accommodated with dominant ideas of development and national interest. As an exception to the norm, the Adivasis have historically dwelled in the margins, recurring only when the political discourse needs to manifest its morality of inclusion and equality. A classic demonstration of this exceptionalisation can be found in the disparate approaches taken by Adivasis and others to the theory of property. The manner in which the theory of property conceives of land is antithetical to how the Adivasi communities have perceived it for centuries (Bijoy, 2008, 2020).

Speaking from within a limited scheme of the dominant and postcolonial language, Adivasis have been accorded merely an insignificant identity –as 'primitive' in their ways of living and belonging particularly to uncivilised societies (Berkhofer, 1979). Along with this positioning, indigenous communities receive a set of characteristics: forest isolation, animism, primitive occupation, illiteracy, carnivorous diet – the list goes on. What could they possibly have to say about 'national' issues?[10] The idea that communities could also possess a history of their own is far from all boundaries of perception; that, communities could synergise themselves with ecosystems and that terms of democracy could be constructed around their distinctive notions of *governance of the self* (Foucault, 2010). The dominant systems and arrangements that have been put in place are "narcissistic self-representations of those who occupy it" (Connolly, 1995: 192), other than of those who support it.[11] The extent of these institutions goes far enough to make Adivasi communities a minority that is to be merely tolerated, since their existence cannot be entirely averted (see Chatterjee, 1986, 1993). Our senses of nationalism[12] and national culture are also diffused with the ideas of a *civilised nation-state* that believes in one idea of democracy (one that believes in majoritarian rule) as a mode of governance (Berger, 1991). Surrounding these ideologies are the minorities, whose space to manoeuvre depends increasingly on the levels of tolerance felt by the nation-state, or more specifically, the majority (Berger, 1991). This represents an integrationist approach to the 'Adivasi problem'. It presumes that the only way Adivasi communities can be included and integrated is by erasing their 'otherness' while ignoring the fact that it is this 'otherness' that underpins Adivasi identity. Adivasis, therefore, are not regarded as a norm of social order, but as communities that by reason of differing find a rare theoretical basis to legitimise their ways of living. Rebuilding this relationship between the majority and minority requires the demolition of dichotomous tendencies and the reconstruction of a relationship that has been, up until now, built on terms of toleration and integration.

In a way, the sovereign reasserts these difficulties to help maintain its dominance. This manner of discourse has consistently made everything else the alternate to the dominant, the other to the norm. Sovereignty remains deeply entrenched in the notion of a singular (and ultimate) rule-making authority within a legal system. It affords a single system of norms, processes, and institutions which reinforces its singularity. In the contest between *formal* and *popular* interpretations of law, it is this overwhelming concept that maps the rules of play. And therefore, any attempt at conceptualising Adivasi assertions needs to begin with displacing sovereignty from its comfortable position of unequivocal power, where it needs to interact with other forms *without asserting dominance*. The next section seeks to add a step in that direction.

Forging Horizontal Relationships

In the depths of political-legal philosophy, there is often only doubt and uncertainty, and perhaps, some theoretical convolutions. Answers are sought to questions that are in themselves conflicted to overcome boundaries of uncertainty and doubts. But in the process, it is often forgotten that, in the crudest sense, we all are sovereigns (Derrida, 2005). We have sovereignty over ourselves, our bodies, and our minds.[13] But an arbitrary distinction is constructed between the sovereignty of the state and the autonomy of individuals and communities. And it is considered *regrettable*, but *necessary* to do so (Shaw, 2008: 39). The practical constraints of polity demand that such a distinction be made but as we have seen before, there is no rational basis for doing so. The nation-state emerges by reinforcing this distinction; by creating an environment for the political to exist within a defined shape – where notionally people delegate their own sovereignties to the state through a consent-based political process for the common good.[14] The processes of law-making, institutional enforcement, and discourse creation all fall within the same bracket. The problem with this narrative is that when the state assumes sovereign powers, it leads to the conceptual demolition of all other forms of powers. It is forgotten that within the political arrangement of a sovereign nation-state, and notwithstanding some degree of power delegation, individual and collective autonomies are not completely extinguished and retain a central position. Indeed, we contend that sovereignty, like any form of power,[15] is characterised by multiplicity and is not a monopoly of the state – that it can and does occur in people and communities beyond the state. Autonomy is what emerges in this process to confront its all-encompassing presence (Foucault, 1977).

It is because both power and authority are concentrated in one monochromatic version of a sovereign-state, that the concept of autonomy is found in direct collision with sovereignty. The concept of autonomy, despite its limited liberalistic origins, challenges sovereignty but often gains very little ground against it (Ivison et al., 2000). Its conceptualisation is limited by the essentials of sovereignty and by the language of rights, where the highest power needs to reside within the state and assertion of rights is completed by the existence of a political. Adivasi communities ascertaining autonomy may do so because it is their *right*, but it is also the *right* of the sovereign state to claim *eminent domain* over their land occupation systems, *ownership* over their resources and *sovereignty* over their autonomy. Adivasi claims, like when they erect stone slabs and interpret the Constitution, are framed and understood within the language of such a theoretical discourse that renders their claims merely incidental to the leading problems of governance.

Under the Indian Constitution, the authority of regimes pre-existing colonisation and the formation of the postcolonial state is claimed against state sovereignty under the language of "autonomy" and "special provisions". Whereas the idea of sovereignty purports an understanding of oneness – one nation, one state, one democracy and one identity – the idea of autonomy, or autonomies, in fact, does quite the opposite. Autonomy, as a concept, seeks to oppose the centralising tendencies of sovereignty when conceived in its monolithic model.

Globally, indigenous people claiming autonomy pre-dating the creation of the State have consistently rejected the hegemony of state sovereignty, which has historically been an instrument of negating their claims over land and resources, as well as their rule-making authority over social, political and economic affairs, and indeed, over knowledge itself (Shaw, 2008). Recognised as 'indigenous peoples' under international law, their assertion of authority over rule-making and governance, particularly over land, resources, and customs, produces tensions for democracy among the multiple culturally determined idioms of 'legality' and 'development' (Pottier et al., 2003; Sachs, 1992; Tucker, 1999). This tension is produced between the ideas of sovereignty as the singularity of political authority within a territory and as an expression of the inherent right to autonomy of peoples within a territory.

In contradistinction with the political of sovereign-state is the other idea that both state sovereignty and autonomy can co-exist, while differing from one another and sometimes even incommensurately so, in the plural. That, sovereignties and autonomies is the more appropriate phrase to represent the phenomenon and that their existence is rather co-dependent. The political process of electing the sovereign can seldom diffuse the sovereignties that the citizens already possess by the very reason of their existence; neither can it undermine the autonomies that both individuals and communities exercise in creating their normative universes (Cover, 1983). The Constitution, understood in its underlying spirit of equality, provides us all, apart from fundamental rights, a wide spectrum of legalities to define and operate our identities. While constructing the sovereign-state model, the Constitution reserves spaces for the public to ascertain their identities independently of that sovereign – whether religious, cultural or political.[16] In fact, the very idea of democracy is founded upon the principle of governing of the self. If the rationale of democracy is to force the state to both find commonalities of identities and recognise the differences in varying communities, then, that has an immediate implication on how the sovereign state model should be organised. Instead of flattening identities in terms of binaries of self and other, inside and outside, civilised and savage, citizen and alien, and modern and primitive, and crystalising a monolithic conception where minority claims need to be accommodated or tolerated, a democratic sovereign society needs to recognise the reality of legal and political pluralism (Cover, 1983; Davies, 2017; Tully, 1994; Walzer, 1983).

Sovereignty and autonomy, both in their singular and plural forms are two distinct forms of power that constantly converse and balance with each other in a functioning democracy. They both co-exist in order to reinforce each other's rationale: the sovereign state derives its legitimacy by promising to protect individual and community autonomies. And different individuals and communities come together to form a sovereign state for the purpose of feeling secure and connected. One without the other not only loses its strength of conception but also the legitimacy of theoretical occupation. The sovereign state is one and popular, but it is with these multiplicities that it gains strength and legitimacy. In this sense, the function of sovereignty is to protect the co-existence of these autonomies and not destroy them. Their relationship, therefore, must be imagined in some form of a horizontal adaptation instead of a hierarchical one.

Communities, especially Adivasis, have exercised, what Cover (1983) calls a *normative mitosis*, for centuries. From time to time, Adivasi communities have operated on the powerful assumptions of being sufficiently autonomous to reinforce their identities through the process of self-governance and private law-making. Communities have taken well-understood and universally accepted devices, like the right to equality, freedom of association, religious freedom, principles of self-governance, directive principles of distributive justice, and resource allocations, to create a radical transformation of perspective in their view of themselves and the outside world. In this event of transforming the perspective, all these legal devices become more than just a rule that is to be literally interpreted; they constitute what we have been calling a normative universe for these communities. The world, for these communities (if we can also add, legitimately so), is turned inside-out and a wall of difference is formed, whose shape differs "depending upon which side of the wall our narratives place us on" (Cover, 1983: 31). Therefore, communities with a total life-vision seek a space in the political structure of the larger society that they are part of. The state's implicit or explicit recognition of their state of being is viewed from their worldviews, which run on presuppositions of autonomies of themselves and sovereignties of the society at large. Adivasi communities are therefore, constantly challenging dominant views of the world and are forcing to rethink the structures formulated to bind and distinguish between knowledges and imaginations – of multiple, coexisting, overlapping, interdependent, and not necessarily incommensurable autonomies, and by implication, legalities.

With the inception of the Indian state, some – but not all – of these imagined systems of governance were incorporated by the Constitution.[17] It structuralised the circulation of these multiple loci of powers, converting them into a range of fundamental rights and self-governance mechanisms. Every region came with its own historicity and culture of governance and the

Constitution endeavoured to protect them all. Specific relationships were established with each region in order to balance the autonomies of the regions with (the claimed) sovereignty of the state. Yet, this was not a simple act of balancing competing interests but a process of reimagining the co-existence of different forces whose boundaries bled into one another. A patchwork of differentiated governance structures was formed and negotiated to produce various levels of constitutional autonomies. As against popular notions, the sovereignty of the Indian state is thus neither monolithic nor a given, but is constituted through a dialectical relationship with the multiple political, social, cultural, and economic autonomies nestled within it. And this relationship needs to be continuously reinforced. Illustrating these different autonomies, the next section demonstrates how sovereignty is shared and how the peoples of India retain their right to interpret the Constitution and the power to govern themselves.

Multiple Autonomies and Shared Sovereignties under the Indian Constitution

The dominant nationalist narrative pegs the birth of the Indian state as an outcome of anti-colonial struggles in the subcontinent, whose shared history and culture provide the "resonance" for the existence of a nation (Shaw, 2008). Article 1 of the Constitution of India, 1950 declares the nation to be a 'Union of States'; a 'union' and not a 'federation' because the states do not have the right to secede. The nation was meant as one indestructible unit, an integral whole whose people were living under a single *imperium* derived from a single source. However, the creation of the Indian state was not a moment, but a process through which peoples within the present territory living under different historical, political, material, and cultural conditions were gradually included in the fold. The Indian Constitution, while constituting sovereign norms and institutions of governance, also recognises these historical negotiations of peoples exercising their political self-determination in joining the Indian Union.

Autonomies as Self-Determination under the Indian Constitution

A structural reading of the constitutional relationships borne through these historical processes of self-determination reveals the existence of multiple and overlapping autonomies constituting 'sovereignty' over land, forests, and resources. Indian constitutional unity is thus not only limited to a romanticised idea of oneness and, by implication, singularity. It is also characterised by a historicity of plurality – of power and of rights. Within this unity is an asymmetrical federalism which creates and protects many different types of autonomies and governance structures (Mahajan, 2007; Rao & Singh, 2004; Saxena, 2012). The Constitution itself demonstrates a patchwork of subjectivities, sometimes overlapping, which nestle different zones of autonomies within the rubric of the Indian state. Several of these are nationalist and sub-nationalist in character, which carry collective imaginations of distinction from others and resonance with each other, based on shared histories and markers of identity such as language, ethnicity, and culture. Equally, each of these zones of autonomy is a culmination of distinct constitutional negotiations between the Indian State and its peoples. Among others, Article 1 and the Fifth and Sixth Schedules constitute this paradoxical co-existence of the one, and the many. In 1996, the Panchayat (Extension to Scheduled Areas) Act extended and enshrined self-determination in the Scheduled Areas by turning the Gram Sabhas into the centre of village self-governance by recognising their various rights and powers. Combined with the 2006 Forest Rights Act, this amounted to a redefinition of governance and democracy under Indian law (Bijoy 2022). The FRA makes a multitude of provisions enshrining forest communities as legitimate authorities over forest conservation and governance. The one Indian State and the many different levels of autonomous units of individuals, states, councils, Gram Sabhas, and communities continuously engage with each other to produce this paradox. Power is distributed by law to all of them and all have the power to interpret the law, to seek to change it, and to govern themselves. It is this horizontal multitude of relationships that forms the one Indian State.

While incorporating these multiple autonomies, the Constitution does not spontaneously extinguish or subsume them in the formation of a larger state but rather protects them on differentiated terms. The liberal democratic frames mostly delimit the 'public' sphere of state action and the 'private' sphere of individual autonomy. The Indian Constitution, on the other hand, demonstrates a different continuum of individual and collective autonomous subjectivities burrowed within the sovereignty of India. These multiple and nestled autonomies manifest as

- i Subject-matter-based distribution of rule-making power. Here, the constitutive autonomous units may retain rule-making authority, delimit the authority of the state through guarantees of individual and collective autonomy under Part III on fundamental rights, or demarcate legislative competence over land, customary law, etc.
- ii Institutional arrangements over rule-making, such as through local self-governance, autonomous councils and associations, customary authorities, as well as representation in federal legislative bodies.

For instance, the Fifth Schedule, as a constitutional instrument recognising the existence of such indigenous autonomies, was the culmination of rebellions and political movements in the colonial and post-independence period for indigenous control over natural resources throughout the 19th and 20th centuries (Bijoy, 2020; Munda & Mallick, 2003; Sundar, 2005). Between 1830 and 1920, Adivasi-dominated eastern states of India witnessed a series of militant rebellions against early attempts of the colonial government to establish eminent domain over indigenous lands and forests through executive and legislative measures, and against native settlers, especially moneylenders. In response, over time, the British colonial government introduced laws recognising the operation of customary laws and institutions, prohibiting land alienation, and 'partially excluding' these areas from British administration. Local governments had the power to determine whether colonial legislations would be extended to these "scheduled districts" or not (Sundar, 2005). After the Independence, the partially excluded areas came to be listed under the Fifth Schedule of the Constitution and the excluded areas of the Northeast came under the Sixth Schedule.[18]

The Constitutional Scheme of Shared Sovereignties over Land and Resources

The Fifth and Sixth Schedules contain differentiated mechanisms for the administration of Tribal Areas. The Fifth Schedule limits the legislative power of the central and state legislatures over these areas and Governors are the constitutional nominees with formal legislative and administrative authority. They embody a constitutional bridge between the central Indian state and the autonomous region. Their obligation is to ensure that the central sovereign power does not engulf the autonomous region and, thereby, to maintain the convergence of the two. They are bound to act in consultation with the Tribes Advisory Council, comprising elected representatives of Scheduled Tribes. Governors have the authority to exclude, modify or limit the applicability of central or state legislations over Scheduled Areas. They are specifically empowered to pass regulations to prohibit or restrict land alienation, to regulate allotment of land, and to regulate money lending. Governors' regulatory powers override central and state legislations on these subjects and are wide enough to even repeal or amend them (Nayak, 1967).

However, existing research shows that these wide powers of Governors are scarcely exercised in practice, and remain largely dormant (Veeresha, 2017). The governance of land, forests, rivers, minerals, and other natural resources are subjects over which the Constitution vests simultaneous authority in the Indian state, as well as in the autonomous Adivasi constituencies. These dot a continuum of autonomy arrangements within the overall constitutional scheme between issues exclusively under state sovereignty and others on which the state has no rule-making authority. This limitation of legislative power is even deeper in Sixth Schedule Areas. Autonomous regions and districts are established in these areas which have direct legislative authority exercised through Regional or District Councils (Nayak, 1967). Governors, as representatives of the central government, have the prerogative to nominate at most four members to the Councils which otherwise comprise about 30 members elected through adult franchise.[19] Once constituted, Councils become autonomous. However, it is to be noted that the power to identify and categorise peoples or areas as Scheduled Tribe/Area or Autonomous District/Region is reserved entirely to the executive machinery of the state, through the President and Governors, such that the exercise of autonomy through self-identification is denied. Said state identification is conducted not on the basis of constitutional rights, but on the condition of deprivation, such that communities are required to establish their economic, educational or social 'backwardness' in order to attain the constitutional rights of autonomy. This acts not only as a potent method of exclusion from constitutional rights but also enables the state to intervene in organic political configurations to disrupt claims and interests.

Nonetheless, the range of multiple autonomies interacting with the authority of the state reflects distinct sources of rule-making authorities. From the Parliament at the centre to the Gram Sabha at the hamlet, the Constitution formulates different rule-making authorities at all levels of governance. The authorities of the state and autonomous units do not operate in water-tight compartments with clear delimitations of power. Instead, the Constitution recognises a spectrum of institutional and subject-matter-based arrangements relating to autonomy and self-determination of Adivasi and indigenous peoples over land and resources and thus opens up a field of political and democratic contestations over rule-making and governance between the two. In this space, where the authorities of the two are deeply enmeshed, tensions between the centralising tendency of a 'singular' sovereignty and the decentralising pulls of its multiple constitutive autonomies are common.

One of the most interesting examples of this constitutional idea of autonomy and of its tensions with singular sovereignty comes from the Niyamgiri Hills located in two scheduled districts of the Adivasi-dominated eastern Indian state of Odisha. In 2003, Vedanta agreed with the state government of Odisha the establishment of a bauxite mine and alumina refinery in the Niyamgiri Hills. For a decade, local Adivasi communities, the Dongria Kondhs, mobilised against the project through many large protests, international campaigns and legal actions. In a landmark 2013 decision over the case,[20] the Supreme Court expanded on the constitutional scheme of autonomy by recognising the right of Gram Sabhas to collectively protect their customary resources, and manage their use and allocation. Following this verdict, the matter was placed before 12 Gram Sabhas who unequivocally rejected the proposed project. However, the conceptual power of sovereignty was felt in the aftermath of this landmark judgement as the central and state governments have persisted in proceeding with the project under the sovereign narratives of 'development' and 'national security'. While protest has continued, human rights violations have also escalated in the area due to a militarisation of the state response (paramilitary activity, security camps, arrests, surveillance, raids, harassment, etc.). These developments reveal the manner in which the state is able to mobilise its defence and security apparatus to override constitutional arrangements of autonomy. State sovereignty has developed in a fashion that empowers it to forge narratives and instruments aimed at gaining control over all existing autonomies. Not only has the state-controlled these autonomies, but has also extinguished their claims to legitimacy. Much like the Pathalgarhi movement, the Adivasi communities of Odisha have been confronted with an all-encompassing state with a sovereign claim so strong that their existence is defined by it.

Concluding Remarks

It is in the hills of Niyamgiri and in the forests of Jharkhand that the contradiction between constitutional arrangements is painfully discernable. Such movements and protests occur as a push-back against an all-encompassing state and become an assertion of the foundational principle of multiplicity. They are an expression of people who were historically promised freedom and autonomy and who have now become the sufferers of sovereignty. They open a path to imagine a decolonisation of sovereignty. Revolutionary poetry has had a significant role in providing a language to articulations that stand at the margins of society. Among the many couplets, poems, and songs fuelling the world, there is one, written by Mirza Ghalib, that aptly describes the narratives coming from the hills of Niyamgiri and the forests of Jharkhand: "*har ek baat pe kahte ho tum ki tū kyā hai, tumhīñ kaho ki ye andāz-e-guftugū kyā hai*" (at every turn you question me, asking 'what are you?' tell me pray what manner of speech do you pursue?).[21]

The predominant and state-led judgement on the pathals and the resulting protests are based on the identity or source of their origins, not their morality, legitimacy or even necessity. Instead of understanding how and why such forms of resistance have emerged, it is preconceived that legal and constitutional interpretations are not the business of the uneducated. The imposition, the arrests, the militarisation, and everything that comes with it, are the violence of sovereignty unleashed on movements questioning the morality of state policy. What is then, this manner of speech pursued by the state and the dominant discourse it dictates that dismisses assertions simply because they originate from people who are presumably unqualified and, therefore, irrelevant?

When we began this chapter with a narration of the Pathalgarhi movement, we wanted to situate the massively complicated questions of sovereignty and its relationship with autonomy. And it was important to do so, for otherwise, the overarching generalising tendencies of the subject would have engulfed us. We saw how the concept of sovereignty has colonised the Adivasi discourse to only offer difficulties in the articulation of any power apart from state sovereignty. We understood that we may have to bend the rules of play, a bit, and unravel the problems they hold within. The problem lies in the exceptionalisation of Adivasi concerns and in the centralising tendencies of sovereignty. Despite the many advantages enjoyed by the concept, we also saw how constitutional jurisprudence nestles, within its bosom, multiple autonomies and multiple sovereignties. And that was the essence of this chapter: to show that there is a way in which sovereignty can be decolonised and read without asserting dominance; that there are multiple versions of these powers theorised differently within the same democratic polity; that they are constructed in a horizontal relationship, not a vertical one. The vertical relationship is popularised by the politics of dominance and in order to have a free intellect and an empathetic disposition, careless adherence to dominance will have to be loosened. The process of decolonising our political imagination and configurations demands the concept of sovereignty to be read outside of its dominant understanding. It also demands a certain creativity of language which we have tried to articulate. What happens to Pathalgarhi and to the Niyamgiri protests will depend, to a considerable extent, on the manner in which we understand the concepts of sovereignty and autonomy and seek to decolonise them; on whether we chose to remain within the manner of speech pursued by the state by the simple virtue of its sovereign powers.

Notes

1 See S. 2(a) 2(1)(i) and 5 of the 2006 FRA. Adivasi people are the indigenous people of India who are designated as Scheduled Tribes under the Constitution. However, there is some political conflict around the term. Hailing from the Hindi language, this term is not accepted by all Indian tribes. Tribes in the Northeast region, for example, prefer to be called 'indigenous communities' and not Adivasi. The term also does not encompass other communities who are categorised under law as "Other Traditional Forest Dwellers". Therefore, the term must be used and understood with the limitations it has. Since this chapter is about the indigenous people of the central Indian belt who are recognised as Adivasi, the term has been used.

2 Gram Sabhas are village assemblies comprising all the adult members of the village. They are constitutional bodies responsible for village administration. The Gram Sabhas have powers to govern their villages and resources under the Constitution and laws such as the 1996 Panchayat (Extension to Schedules) Areas Act and the 2006 Scheduled Tribes and Other Traditional Forest Dwellers (Recognition of Forest Rights) Act.

3 The 1950 Fifth Schedule of the Constitution of India refers to a constitutional scheme of autonomy and self-governance of customary homelands of Adivasis, or indigenous peoples, which recognises the right to self-governance through customary laws and institutions. These areas are administratively demarcated as per the process laid out in the Constitution itself.

4 This is a translation of the writing on stone slabs at a village in Khunti, Jharkhand.

5 Munda community is a dominant Adivasi community living primarily in the Chota Nagpur region of Jharkhand. However, they are also found in other parts of Jharkhand, Odisha, West Bengal, and adjacent areas of Chhattisgarh.

6 Jal, that is water, jangal, that is forest, and jameen, that is land is one of the most popular articulations of indigenous autonomy in India. The first villages to declare Pathalgarhi in Jharkhand in 2017, spread over the area of Khunti, were the birthplace of Birsa Munda and the site of militant rebellions against the British colonial powers in the late 19th and early 20th centuries. As the British government sought to impose colonial dominion over forest land and resources by extinguishing customary rights and traditional systems of forest governance through legal and militaristic means, the rebellions were instrumental in the enactment of the Santhal Parghana Tenancy Act (SPTA) in 1855 and the Chhotanagpur Tenancy Act (CNTA) in 1907. Both CNTA and SPTA preserved Adivasi autonomy and self-governance over *jal*, *jangal*, and *zameen* to be exercised through traditional Gram Sabhas without intervention by the British government, and deemed these lands as inalienable to Adivasis. As the colonial project of resource appropriation extended across colonial India in the succeeding decades, specifically encoded by the 1901 Land Acquisition Act, 1927 Indian Forest Act, and a range of other laws relating to land and natural resources, CNTA and SPTA continued to protect Adivasi autonomy over homelands well into post-Independence India and the introduction of the Fifth Schedule to the Indian Constitution.

7 For instance, in many parts of Jharkhand including Latehar and Palamu, villages declared self-governance under the auspices of the Panchayat (Extension to Scheduled Areas) Act (PESA) in 1996 through the erection of similar stone slabs declaring their rights and powers of self-governance under the Indian Constitution and statutory law.

8 The concept of sovereignty has been theoretically strengthened over the years. Thomas Hobbes ([1651]1968) was one of the earlier theorists that strongly propounded the concept. Some of the other phenomenal works include Schmitt (1985), Bartelson (1995, 2001), and Grimm (2015).

9 A lot of writing on the subject, especially the one post-FRA-PESA era, quite rightly, suggest for a paradigm shift in the thinking of Adivasi land rights. Although these two enactments have recreated, to a considerable extent, the law relating to Adivasi land rights leaving a lot on the implementation mechanism, but, as this essay will argue, the organisation of the discourse itself needs a certain amount of rethinking.

10 See Thakkar Sub-Committee (1947) for an instance of this kind of thinking and argumentation.

11 The instability of the centre is now covered by the narcissistic self-representations of those who occupy it. To the extent such a drive exceeds, every other interest, faith and moral orientation now becomes a minority to be tolerated or corrected by the self-occupants of the centre. The national culture becomes one with an unmarked constituency at the centre, surrounded by various minorities whose space to manoeuvre depends increasingly upon levels of tolerance or intolerance felt by the unmarked constituency. These minorities – sometimes numerically enough to be numerical majority are now set-up to become objects of vilification, discipline, regulation and violence when things go wrong anywhere in the state.

(Connolly, 1995: 192–3)

12 For an understanding of nationalism and its connection with power, see Orwell (1945).

13 While non-sovereignty theorists have questioned and nuanced the liberal modern understanding of self-sovereignty (Berlant, 2011; Berlant & Edelman, 2014), sovereignty over self is still an important ideology of how one conceives of one's body and mind and their association with others.

14 The theory of social contract is concerned with investigating the legitimacy of the authority of the state over an individual. The most contemporary articulation of the concept was suggested by John Rawls ([1971] 2005).

15 Michel Foucault (1980) has been influential in shaping the understanding of the concept of power. His famous phrase "power is everywhere" and "comes from everywhere" signified that power was not merely restricted in actors who use power as an instrument of coercion and the discreet structures in which they operate but was spread across the political spectrum.

16 The Indian Constitution reserves multiple spaces to preserve individual and community autonomies. Part III enlists a range of fundamental rights that individuals and in some cases, communities can exercise as their power against the state. There are detailed provisions for rights to freedom of religion in Articles 26–28 and cultural rights in Articles 29 and 30. Part IX provides for structures of self-governance. Both Panchayat and Gram Sabha are constitutional institutions that are empowered to govern at the local level. Part X, particularly Articles 244 and 244A read with Schedules V and VI are special provisions to set the governance mechanisms in tribal areas.

17 Adivasis and other tribal communities were hardly represented in the Constitution-making process. Protections were compromised to a large extent as Jaipal Singh Munda's concerns and suggestions, the only person talking about this matter, were not all taken into account. Therefore, the process of constituting these autonomies was also compromised. But even within the existing structure, there is ample space for interpretation and reimagination, to which this chapter intends to contribute.

18 The history of the Northeast is diverse and out of the scope of this chapter.
19 Proviso to Section 2, Schedule VI Constitution of India 1950.
20 See Odisha Mining Corporation v. Ministry of Environment, Forests and Others (2013) 6 SCR 881.
21 This is a couplet by the famous Urdu Poet Mirza Ghalib. Full poem with translation available at https://www.rekhta.org/ghazals/har-ek-baat-pe-kahte-ho-tum-ki-tuu-kyaa-hai-mirza-ghalib-ghazals

References

Bartelson, J. (1995). *A Genealogy of Sovereignty*. Cambridge: Cambridge University Press.

Bartelson, J. (2001). *The Critique of the State*. Cambridge: Cambridge University Press.

Berger, T. (1991). *A Long and Terrible Shadow: White Values, Native Rights in the Americas, 1492–1992*. Seattle: University of Washington Press.

Berkhofer, R. F. (1979). *The White Man's Indian*. New York: Vintage Books.

Berlant, L. (2011). *Cruel Optimism*. Durham: Duke University Press.

Berlant, L., & Edelman, L. (2014). *Sex, or the Unbearable*. Durham: Duke University Press.

Bijoy, C. R. (2008). Forest Rights Struggle: The Adivasis Now Await a New Settlement. *American Behavioral Scientist*, *51*(12), 1755–1773.

Bijoy, C. R. (2020). The Forest Rights Struggle and Redefining the Frontiers of Governance: Dismantling hegemony, Reconstructing Authority and Collectivizing Control. In V. Bhagat-Ganguly & S. Kumar (Eds.), *India's Scheduled Areas: Untangling Governance, Law and Politics* (pp. 75–100). London: Routledge.

Bijoy, C. R. (2022). Commentary: Tracing Panchayati Raj's Role in Deepening Democracy and Decolonising Governance Over 30 Years. *Mongabay*, 25 May 2022. Availableat:https://india.mongabay.com/2022/05/commentary-tracing-panchayati-rajs-role-in-deepening-democracy-and-decolonising-governance-over-30-years/

CDRO-WSS (Coordination of Democratic Rights Organization and Women Against Sexual Violence and State Repression). Fact Finding of Khunti, Ghagra, Palamu Tiger Reserve and Sedition Cases. 21 August 2018. Available at: https://wssnet.org/2018/08/21/press-release-of-the-cdro-and-wss-fact-finding-of-khunti-ghagra-palamu-tiger-reserve-and-sedition-cases/

Connolly, W. (1995). *The Ethos of Pluralization*. Minneapolis: University of Minnesota Press.

Chatterjee, P. (1986). *Nationalist Thought and the Colonial World: A Derivative Discourse?* London: Zed Books.

Chatterjee, P. (1993). *The Nation and its Fragments: Colonial and Postcolonial Histories*. Princeton: Princeton University Press.

Cover, R. (1983). Foreword: Nomos and Narrative. *Harvard Law Review*, *97*(4), 4–68.

Davies, M. (2017). *Law Unlimited: Materialism, Pluralism and Legal Theory*. London: Routledge.

Derrida, J. (2005). *Sovereignties in Question*. New York: Fordham University Press.

Dussel, E. (1995). Eurocentrism and Modernity (Introduction to the Frankfurt Lectures). In J. Beverley, M. Aronna, & J. Oviedo (Eds.), *The Postmodernism Debate in Latin America* (pp. 65–77). Durham: Duke University Press.

EPW Engage (2019). Jal, Jangal Aur Jameen: The Pathalgadi Movement and Adivasi Rights, *Economic & Political Weekly*, 14 October, Available at: https://www.epw.in/engage/article/pathalgadi-movement-nation-autonomy-rights-adivasi-jharkhand

Foucault, M. (1977). *Discipline and Punish: The Birth of the Prison*. New York: Vintage Books.
Foucault, M. (1980). *Power/Knowledge: Select Interviews 1972–1977*. New York: Pantheon Books.
Foucault, M. (2010). *The Government of the Self and Others: Lectures at the College De France 1982–83*. London: Palgrave McMillan.
Grimm, D. (2015). *Sovereignty: The Origin and Future of a Political and Legal Concept*. New York: Columbia University Press.
Guha, R. (1998). *Dominance without Hegemony?: History and Power in Colonial India*. Cambridge: Harvard University Press.
Guha, R. (2017). *Democrats and Dissenters*. New York: Penguin Books.
Hobbes, T. ([1651] 1968). *Leviathan*. New York: Penguin Classics.
Ivison, D., Patton, P., & Sanders, W. (Eds.). (2000). *Political Theory and the Rights of Indigenous Peoples*. Cambridge, MA: Cambridge University Press.
Mahajan, G. (2007). Federal Accommodation of Ethnocultural Identities in India. In B. He, B. Galligan, & T. Inoguchi (Eds.), *Federalism in Asia* (pp. 82–100). Cheltenham: Edward Elgar.
Maldonado-Torres, N., Vizcaíno, R., Wallace, J., & We, A. J. E. (2018). Decolonising Philosophy. In G. Bhambra, D. Gebrial, & K. Nişancıoğlu (Eds.), *Decolonising the University* (pp. 64–90). London: Pluto Press.
Mignolo, W. (2007). Delinking: The Rhetoric of Modernity, the Logic of Coloniality and the Grammar of De-Coloniality, *Cultural Studies*, *21*(2), 449–514.
Monahan, M. (2011). *The Creolizing Subject: Race, Reason, and the Politics of Purity*. New York: Fordham University Press.
Munda, R. D., & Mallick, S. (2003). *The Jharkhand Movement: Indigenous Peoples' Struggle For Autonomy In India*. Copenhagen: IWGIA.
Nayak, R. (1967). Powers of the Governor under Sixth Schedule of the Indian Constitution. *Journal of Indian Law Insititute*, *9*(2), 237–245.
Orwell, G. (1945). *Notes on Nationalism*, London: Penguin Modern.
Pottier, J., Bicker, A., & Sillitoe, P. (Eds.) (2003). *Negotiating Local Knowledge: Power and Identity in Development*. London: Pluto Press.
Quijano, A. (2007). Coloniality and Modernity/Rationality, *Cultural Studies*, *21*(2), 168–178.
Rao, G., & Singh, N. (2004). *Asymmetric Federalism in India*. UC Santa Cruz: Department of Economics, UCSC. Retrieved from https://escholarship.org/uc/item/0v59942g
Rawls, J. ([1971] 2005). *A Theory of Justice*. Cambridge: Belknap Press.
Rights and Resource Initiative (RRI); Vasundhara & Natural Resources Management Consultants (2015) *Potential for Recognition of Community Forest Resource Rights Under India's Forest Rights Act*. Available at: https://rightsandresources.org/publication/potential-for-recognition-of-community-forest-resource-rights-under-indias-forest-rights-act/
Roth, D., & Mullins, C. (2010). The Death of State Sovereignty? An Empirical Exploration, *International Journal of Comparative and Applied Criminal Justice*, *34*(1), 79–96.
Sachs, W. (Ed.) (1992). *The Development Dictionary: A Guide to Knowledge as Power*. London: Zed Book.
Sarin, M., & Springate-Baginsk, O. (2010). *India's Forest Rights Act - The Anatomy of a Necessary But Not Sufficient Institutional Reform*. IPPG Discussion Papers Series No. 45. Available at: https://assets.publishing.service.gov.uk/media/57a08b0be5274a27b2000909/dp45.pdf

Saxena, R. (2012). Is India a Case of Asymmetric Federalism?, *Economic and Political Weekly*, *47*(2), 70–75.

Schmitt, C. (1985). *Political Theology: Four Chapters on the concept of Sovereignty*. Chicago: University of Chicago Press.

Sharma, S. (2019) 10,000 People Charged with Sedition in One Jharkhand District. What Does Democracy Mean Here?, *Scroll*, 19 November. Available at: https://scroll.in/article/944116/10000-people-charged-with-sedition-in-one-jharkhand-district-what-does-democracy-mean-here

Shaw, K. (2008). *Indigeneity and Political Theory: Sovereignty and the Limits of the Political*. London and New York: Routledge.

Sundar, N. (2005). Laws, Policies, Practices in Jharkhand, *Economic & Political Weekly*, *40*(41), 4459–4462.

Thakkar Sub-Committee (1947). *Report of the Sub-Committee on Excluded and Partially Excluded Areas (Other than Assam)*. Presented before the Constitutent Assembly of India.

Tucker, V. (1999). The Myth of Development: A Critique of Eurocentric Discourse. In R. Munck & D. O'Hearn (Eds.), *Critical Development Theory: Contributions to a New Paradigm* (pp. 1–2). London: Zed Books.

Tully, J. (1994). *Philosophy in an Age of Pluralism*. Cambridge, UK: Cambridge University Press.

Veeresha, N. (2017). Governance of the Fifth Schedule Areas: Role of Governor, *Indian Journal of Public Administration*, *63*(3), 444–455.

Walzer, M. (1983). *Spheres of Justice: A Defense of Pluralism and Equality*. New York: Basic Books.

Wright, S. (2016). Nation Building in the Wake of Colonialism: Old Concepts in New Settings. In S. Wright (Ed.), *Language Policy and Language Planning* (pp.78–111). London: Palgrave Macmillan.

8 Indigeneity, Autochthony, and Belonging

Conceptual Ambiguity as an Impediment to Decolonisation in South Africa

Rafael Verbuyst

Introduction

On 18 March 2022, a South African judge delivered an unprecedented judgement: "the fundamental rights of First Nations Peoples" were insufficiently considered in the context of a multi-billion-rand property development in Cape Town (*Observatory Civic Association et al. v Trustees for the time being of Liesbeek Leisure Properties Trust et al.* [2022] 12994/21: 77–9). Construction was halted pending greater consultation with "Indigenous Groups, more particularly the Khoi and San First Nations Peoples". What makes this injunction remarkable is not so much that it acknowledges the Khoi and San's (Khoisan[1]) undisputed historical footprint in the area, but that it recognises their indigeneity and "fundamental rights" – although, notably, among other unspecified "Indigenous Groups" – in a legal setting, something which the post-apartheid dispensation has refused thus far.[2] Where the South African state is concerned, there are no 'indigenous people' within its borders. In fact, the 2019 Traditional and Khoi-San Leadership Act, which stipulates the criteria for the recognition of Khoisan and non-Khoisan traditional leadership, unequivocally rejects "any special indigenous, first nation or any other similar status" under Section 1(1)(*2*). Confusingly, President Cyril Ramaphosa (2019) nevertheless celebrated the Act for granting "statutory recognition to the Khoi-San [...] one of South Africa's indigenous groups". Equally jarring was South Africa's signing of the 2007 United Nations Declaration on the Rights of Indigenous Peoples (UNDRIP), which recognises indigenous rights in terms of land, culture, and self-determination, as this was not accompanied by an endorsement of the United Nations' long-time recognition of the Khoisan as indigenous people (Stavenhagen, 2005). The UNDRIP is non-binding and needs to be translated into national laws to be effective, but this did not happen and its primary stakeholders remain unspecified.

These examples reflect a pattern that has emerged after the end of apartheid, marked by widespread ambiguity surrounding the term indigenous, whether the Khoisan should be recognised as such, and what such a recognition might entail (Verbuyst, 2022: 98–145). This partially stems from the distinct colonisation of the Khoisan, as well as the nature of their resurgence in the post-apartheid era. The Khoisan's colonisation started in the mid-17th century and

DOI: 10.4324/9781003293460-12

entailed the loss of land, modes of subsistence, and material destruction, but equally loss of language, epistemology, and sense of self (Bam, 2021). Bantu-speaking groups historically settled in the Eastern half of the country, experienced colonialism later, and were ultimately classified 'Black' by the apartheid dispensation. In contrast, the Khoisan, alongside various others, including enslaved individuals from Asia and East Africa, were branded 'Coloured': a mixed-race label denoting lineages that were deemed neither 'White', 'Indian', or 'Black'.[3] 'Coloureds' fared better than 'Blacks' in many respects, but far worse than 'Whites'. Centuries of forceful assimilation culminated in the commonly held notion that the Khoisan are virtually extinct, which was over time internalised by most, though certainly not all, Khoisan as well. Given that racial labels also endure in a semi-official capacity (see below), most people with Khoisan ancestry regard themselves as Coloured today. And yet, a growing number of mostly urban-based activists are rejecting the Coloured label, self-identifying as indigenous,[4] 'reviving' Khoisan culture, and campaigning for recognition and reparations in the post-apartheid era (Verbuyst, 2022). These efforts have catapulted Khoisan issues into the public sphere, resulting in greater awareness about the Khoisan's predicaments, as well as modest political gains, but the conceptual ambiguity remains, if not increases: the Khoisan are simultaneously recognised and unrecognised, their indigeneity acknowledged and denied, and their contemporary and historical marginalisation addressed and ignored.

Using the term indigenous gratuitously diminishes its credibility and ability to pinpoint precise forms of marginalisation (Niezen, 2009: 178–9). Indeed, as I will show, various actors mobilise the concept of indigeneity – and its related terms, 'First Nations' in particular – in vastly different ways. This not only causes misunderstandings, frustrations, and people to talk past one another but also confuses what and who 'decolonisation' is for: who is 'indigenous' and what do they want? Which colonial legacies need addressing? How is this to be accomplished and at whose, if anyone's, expense? Why do some endorse and others reject indigeneity? 'Decolonisation' is understood here as dismantling 'coloniality', or "long-standing patterns of power that emerged as a result of colonialism, but that define culture, labour, intersubjectivity relations, and knowledge production well beyond the strict limits of colonial administrations" (Ndlovu-Gatsheni, 2013: 13). South African coloniality thus manifests itself in enduring socio-economic inequalities along racial lines (Francis & Webster, 2019), but equally on the epistemological and ontological plane. The 2015 student-led 'Rhodes Must Fall' campaign slamming, among others, the University of Cape Town's lingering Eurocentricity, made this highly apparent. Significantly, however, while the protesters were praised for probing beyond apartheid and centring Afrocentric knowledge, they replicated the common blanket binary 'Black'/'White' in the process and hardly acknowledged that the university campus lies on ancestral Khoisan land (Bam-Hutchison, 2016: 12). This glaring omission of the Khoisan and their distinct forms of marginalisation exemplifies exclusionary dimensions of the prevailing discourse on decolonisation, as well as the fact that over three decades of Khoisan activism have not met their goals.

Can 'indigeneity' act as an effective political concept to dismantle coloniality and address the Khoisan's grievances in the face of all this conceptual conflation?[5] I explore this question in different steps: differentiating between currently prevailing uses of indigeneity in practice, policy, and analysis; reframing them as "conflicting logics of autochthony" (Zenker, 2021); and appraising the potential implications for decolonisation. Rather than taking indigeneity's meaning or political implications for granted, one indeed first needs to disentangle different strands of thought on indigeneity among the most impactful groups: the Khoisan, government officials, and academics. No longer associated with its original colonial connotations of underdevelopment and backwardness (Niezen, 2009: 178), 'indigenous' is since the 1980s mostly understood as an international legal term for a people who self-identifies as indigenous, occupied a certain territory before dominant others (so-called prior occupancy), practices a distinct culture, and experiences past and present-day marginalisation. However, as indigeneity's ascriptions vary widely in practice, I will borrow from articulation theory, which examines the unmaking and making of connections between elements that do not necessarily correspond within a given discourse (Verbuyst, 2022: 20–4). Studying 'articulations of indigeneity' then amounts to scrutinising indigeneity's production "through processual, multi-actor, multiscale networks and within specific grounded contexts, each with particular configurations of colonial histories, postcolonial modernities, epistemological-ontological commitments, and formulations of difference" (Radcliffe, 2018: 2).

In what follows, I draw on data collected during several bouts of ethnographic fieldwork among Khoisan activists in Cape Town between 2014 and 2022 (Verbuyst, 2016, 2022), complemented with scholarly critiques, governmental documents, and media items, to distinguish 'etic-critical', 'etic-analytical', and 'emic' articulations of indigeneity. Indigeneity is a self-identifier, but can also denote others – the 'emic' and 'etic' perspectives, respectively. It is important to note upfront that my framework constitutes an analytical abstraction, it does not reflect a moral hierarchy, nor do the different articulations of indigeneity neatly correspond with the groups under review. Moreover, although the Khoisan are by far the most prominent ones, they are not alone in identifying as indigenous in South Africa, which somewhat nuances my use of 'emic' in this text (see below). Lastly, I mostly draw on examples from the specific case of Cape Town-based Khoisan activism, thereby partially reflecting local concerns. However, my broader theoretical arguments are deliberately applied to the South African context as a whole. While each articulation calls attention to different facets of indigeneity, they are conflated by all stakeholders involved, adding to the aforementioned stinted decolonisation process. In the penultimate section, I reframe this apparent dissonance as "conflicting logics of autochthony" (Zenker, 2021) by introducing the concepts of 'belonging' and 'autochthony'. This refurbished analytical toolbox is not meant to resolve differences of opinion, but rather to delineate competing positions within the politics of indigeneity more clearly. Crucially, however, as I argue in the concluding section, this framework does identify marginalised perspectives, as well as productive ways of moving decolonisation forward.

Etic-critical Articulations of Indigeneity

Post-apartheid South Africa rebranded itself as a 'Rainbow Nation', celebrating diversity and granting equal rights to citizens under its new Constitution. Although the government has never spelt out an official position on indigeneity and habitually contradicts itself on the subject, some of its representatives clearly view indigeneity as antithetical to nation-building. In 2013, the Secretary-General of the ruling African National Congress (ANC), for instance, pleaded with Khoisan activists to cease thinking like a "minority" because they are part of the "majority" (as quoted in Verbuyst, 2016: 87). In terms of historical justice, the 1997 White Paper on Land Reform (DLA, 1997: 77–8) dismissed land claims going back further in time than 1913 – i.e., when most Khoisan were dispossessed – as these might provoke "competing claims [which] awaken and/or prolong destructive ethnic and racial politics". These reservations relate to indigeneity's connotations of self-determination, secession, and potential to stimulate competitive victimhood (Kuper, 2003). To tamper such aspirations, some officials have suggested that all Africans are 'indigenous' to South Africa as they are all victims of European colonialism; rendering 'indigenous' synonymous with 'African' (Ndlovu-Gatsheni, 2009: 72–3). Kumarakulasingam and Ngcoya concur that triaging "indigenous co-sufferers" is arbitrary: "A linear understanding of indigeneity does not contend with the crooked lines of subordination and liberation" (2016: 846, 859). Indigeneity, they contend, is "complex and awkward in South Africa", unlike in North America or New Zealand where societal cleavages are ostensibly more clear-cut. Another critic does concede that the Khoisan face unique challenges, but similarly maintains that indigenous status is unjustified given the country's "racially polarised history", as well as the fact that most South Africans share the Khoisan's socio-economic marginalisation (Lehmann, 2004: 109–10).

Echoing common critiques by academics and African government officials that 'aborigine', 'First Nation', and 'indigenous people' are foreign concepts that are misguidedly applied to Africa (Pelican, 2009), a leading South African official opposed "first-nation status" for the Khoisan on the grounds that it might encourage secession and, in contrast to Latin America "where people were completely removed from their land and some arrived and settled there", in South Africa "we do not know who arrived at which point, when and where" (as quoted in Makinana, 2016). The Khoisan, he adds, avowedly shared land willingly with "other indigenous people of Africa". The debate over prior occupancy – 'who' arrived 'where' first – is complex, politically charged, and pertains to nomenclature, descent, inter-ethnic group relations, and contested archaeological evidence (see Verbuyst, 2022: 17–20, 111–4, 135–6). Mellet argues that exclusionary agendas and colonialist pseudo-history – the myth that Bantu-speaking populations ("Black invaders") arrived in South Africa at the same time as Europeans – regularly underpins claims of prior occupancy (as quoted in "Stop Calling us Coloured" 2018; see also Mellet, 2010: 244). He also discerns colonial legacies in certain activists' use of "feudal terminology", such as 'King', and the manner in which they reproduce "the primitivist paradigm of the 'Nobel Native'" by appropriating common stereotypes regarding indigenous people (Mellet, 2010: 29, 225, 244; see also Kuper, 2003). Particularly concerning are claims of (pure) descent on the basis of supposed genetic 'evidence' (Erasmus, 2013). Linking contemporary Khoisan to their historic descendants is complicated – though certainly not disavowed – by various factors, not least that ethnic groups intermingled, causing the genetic strain most commonly associated with the Khoisan to be prevalent in particular among Coloureds, but also Bantu-speaking groups, among others (Barbieri et al., 2014). Arguments that link 'degrees' of genetic strains to qualifications for indigeneity stem from a grave and potentially dangerous misunderstanding of genetic science.

Mellet's work indeed pushes back against what he regards as perilous colonial constructs of purity, primitiveness, and prior occupancy by celebrating diversity, migration, and ethnic hybridity in 'Coloured' history due to the many lineages that have historically contributed to this group. To many, Khoisan activism runs anathema to a "non-racial" South Africa by reinventing pro-segregationist Coloured identity politics through an essentialist and exclusionary emphasis on race, indigeneity, and ethnicity (Besten, 2006; Jacobs & Levenson, 2018). These criticisms reflect the arguably still dominant paradigm in South African academia, which rejects such notions as intellectually defunct, politically dangerous, and opportunistic covers to obtain power and resources (Rassool, 2019). For a long time, this critical reading of Khoisan activism indeed ran virtually unopposed. Critics have legitimate concerns, as well as empirical evidence to back these up, but they reduce indigeneity to notions of purity and prior occupancy, which leaves other approaches largely out of their purview, not least those of the Khoisan themselves. In the following two sections, I address two types of articulations of indigeneity that are currently underrepresented in debates on indigeneity and decolonisation. Before discussing emic perspectives, I turn to a settler-colonial studies' reading of indigeneity, which I, together with Veracini, earlier laid the foundations for (Veracini & Verbuyst, 2020).

Etic-analytical Articulations of Indigeneity

When people speak of South African settler-colonialism, they commonly refer to periods of Dutch, British, and Afrikaner domination. However, the permanent presence of settlers sustains a distinct structure of domination whereby 'settlers' benefit through a "logic of elimination" at the expense of 'natives' (i.e., indigenes), whom they constantly seek to dispossess, contain, and strip of their indigeneity (Wolfe, 2006; Veracini, 2010). Settler-colonialism thus differs from 'colonialism', which is primarily about exploitation rather than elimination and formally ceases when colonialists return to the metropole and formal independence is achieved. Manifestations and justifications of settler-colonialism change over time, possibly even including those who *structurally* benefit as 'settlers' atop of the "population economy" (Veracini, 2010). This change in structural positioning makes naming contemporary South African settler-colonialism highly controversial for it implies that all non-Khoisan, including the Bantu-speaking majority, continue to benefit as settlers, albeit to varying degrees. It might also suggest that some South Africans suffered 'more' than others. While these reservations might explain why contemporary settler-colonialism in South Africa has hardly been explored (see Cavanagh, 2013), they stem from erroneous interpretations of a settler-colonial studies framework. Rather than referring to tangible 'groups' and 'individuals', who rarely fit unproblematically within analytical models or conform to the clear-cut binaries 'settler'/'native', a settler-colonial studies' approach refers to these in the abstract sense as it is primarily concerned with settler-colonialism as a *structure* of past and present domination.

An analysis along these lines is indeed premised on the notion that the Khoisan were colonised differently than other Africans. From the 19th century onwards, Bantu-speaking groups were generally speaking colonised through 'race' for purposes of 'exploitation', whereas the Khoisan were slated for 'elimination' since the mid-17th century. Bantu-speakers were dispossessed and brutalised by settlers, but, crucially, unlike the Khoisan, they did not experience genocide and (the same degree of) forceful assimilation (Adhikari, 2011). During apartheid, the Khoisan were, for instance, not afforded a 'homeland', or indeed 'Bantustan', on the grounds that, contrary to Bantu-speaking groups, whose culture needed to be protected and nurtured, the Khoisan were believed to have been wiped out. This prototypical settler-colonial trope of the "vanishing native" (Brantlinger, 2013) was bolstered by the aforementioned assimilation of the Khoisan as 'Coloured', which acted as an internalised marker of miscegenation and non-indigeneity. The anti-apartheid movement advocated for the liberation of all Africans from White-minority rule, thereby extending the emphasis on 'race', if for vastly different purposes. Similarly, the ANC government implemented a series of policies to address *racial* inequalities after the end of apartheid. Coloureds benefited from these too, although the question of indigeneity was left out. The post-apartheid government's policies remained blind to settler-colonialism, perhaps even (unwittingly) entrenching it further. 'Race' continues to obfuscate 'indigeneity'.

Examples abound, but I will stick to two interrelated manifestations of contemporary South African settler-colonialism. The aforementioned suggestion that all Africans are 'indigenous' because they equally suffered under colonialism certainly comes to mind as a potential instance of "colonial equivocation", or the homogenisation of "various experiences of oppression as colonization" (Tuck & Yang, 2012: 17). This articulation of indigeneity rejects treating the Khoisan differently since they are just as indigenous as other African ethnic groups. Colonial equivocation chimes well with another feature of settler-colonialism: "settler nativity", whereby settlers claim indigeneity to cloud their structurally beneficial position (Tuck & Yang, 2012: 10). Different motivations underpin assertions of settler nativity. Settler (2010), for instance, observed how non-Khoisan traditional leaders increasingly seek to bolster their position in South African society by tapping into the international discourse on indigenous people and the "rhetoric of the ancestral". Then-President Thabo Mbeki similarly, if more subtly, appropriated Khoisan indigeneity in the context of the reburial of Sarah Baartman, a 19th-century Khoisan woman who was paraded in Europe and whose remains were kept in a French museum until 2002. During a speech at the reburial ceremony, Mbeki did not mention the Khoisan but emphasised how Baartman's treatment reflected African women's experiences and dispossession in general (Besten, 2006: 339–42).

South African officials have voiced other articulations of indigeneity, which could in turn arguably be regarded as "deep colonising" measures: ostensibly decolonising settler-colonialism, but in effect anchoring it more firmly (Rose, 1996; Veracini, 2011). One example is qualifying the Khoisan as "first", "vulnerable", or "marginalised" indigenous people – the Khoisan as one of South Africa's indigenous peoples, but suffering from particular forms of marginalisation, which are usually unspecified (see, e.g., SAHRC, 1999; Zuma, 2001). These specific designations have to date not informed any policies that address the Khoisan's grievances and their vague formulation continues to skirt around the thorny issue of what defines the past and present indigeneity of the Khoisan. Deep colonisation is also discernible in land reform. The South African government announced in 2013 that it would re-open the land claims process in part because the Khoisan had been unjustly left out of the restitution program of the mid-1990s (Zuma, 2013). This was due to the constitutionally enshrined cut-off date of 1913 in reference to the Natives Land Act, which assigned most of South African territory to Whites, long after the Khoisan had already been dispossessed (Cavanagh, 2012). However, the 2013 Restitution of Land Rights Amendment Act made no mention of the Khoisan. Promises were made to explore "exceptions" to the cut-off date, but the two national workshops that took place in the Act's aftermath made little headway in this regard (see Verbuyst, 2016). In fact, government officials reached the same conclusion as almost two decades earlier: rather than restitution, redistribution of state-owned land would be preferable in the case of the Khoisan (DLA, 1997). To date, however, hardly any such land redistributions have occurred (see also Huizenga, 2018).

In this section, I opted to discuss 'indigeneity' as an abstract concept in relation to settler-colonialism as a structure of domination, not in reference to specific groups of people. What remains therefore is to turn towards Khoisan activists and their emic articulations of indigeneity. This last step is vital, not least because their perspectives are least of all represented in debates on indigeneity and decolonisation.

Emic Articulations of Indigeneity

Khoisan activists began to navigate international fora in the 1990s and 2000s, swiftly resulting in their recognition as indigenous people by the United

Nations (Stavenhagen, 2005). The identification as indigenous people was intended to add weight to campaigns regarding land restitution, socio-economic marginalisation, and cultural rejuvenation. One of the most influential organisations at this stage was the Cape Cultural Heritage Development Council (CCHDC), a non-profit organisation aimed at getting "historically coloured people" to celebrate their Khoisan roots (Besten, 2006: 288). The CCHDC also criticised affirmative action policies intent on undoing racially skewed employment patterns inherited from the apartheid era. This opposition to affirmative action, which remains in place to date, illustrates the paradoxical simultaneous reification and rejection of Coloured identity by Khoisan activists. According to the CCHDC and others, affirmative action in practice unjustly privileges those who were labelled 'Black' under apartheid, even if there are also provisions for 'Coloureds'. The Khoisan community newspaper *Eerste Nasie Nuus* [First Nation News] is replete with such complaints, opining that the government only cares about uplifting Blacks and that indigenous people are explicitly discouraged to apply for certain positions (see, e.g., Online Readers React, 2013). Others reject affirmative action policies because they require the Khoisan to identify as 'Coloured' in order to qualify for potential benefits (Langeveldt, 2016: 85).

Unsurprisingly, most Khoisan activists reject 'Black' or 'Coloured' identity. However, given their historical entanglement with the latter, they inevitably articulate their grievances in relation to personal and collective experiences of being known as 'Coloured'. This history impinges on the present in many ways. Despite long-time pleas from activists, 'Khoisan' is not an option in the census, but 'Coloured' remains. In (semi-)official settings, 'African' is still frequently used interchangeably with 'Black'. While many veteran Khoisan activists were part of the Black Consciousness and anti-apartheid movements, where 'Black' meant *all* marginalised Africans, they – and others (see, e.g., Ndlovu-Gatsheni, 2009: 65) – feel its meaning has gradually shifted in the post-apartheid era to exclusively denote the Bantu-speaking majority. Irrespective of this, as one Khoisan activist put it, "the term Black does not explore the depth of colonialism" (Verbuyst, 2022: 207). An exclusive interpretation of 'African' is unacceptable to the Khoisan, who seek to accentuate their African roots by emphasising their indigeneity. In doing so, they often face ridicule and doubt, or even insults such as '*amalawu*': a person without a soul, lacking a 'real' history, let alone claim to indigeneity. Conversely, the rejection of Black identity by Khoisan activists is at times embedded in a broader racialised discourse reminiscent of the apartheid era, where Black South Africans are cast as land-grabbing 'foreigners' who disproportionally benefit from social welfare (see, e.g., Jacobs & Levenson, 2018).

Imperilled by 'foreigners', some Khoisan regard secession as their only recourse (Kamaldien, 2018). In 2017, "King Goab Khoebaha Calvin Cornelius III", for instance, claimed to have seceded from South Africa by founding the "Sovereign State of Good Hope"; actions he justified by referring to South Africa's signing of the UNDRIP, but which were ignored by the South African government and widely condemned in the wider Khoisan community (Besent, 2018). Khoisan activist Lesle Jansen, who is part of the Working Group of Indigenous Populations/Communities of Africa of the African Commission on Human and Peoples Rights, stressed that secession or 'being first' in (South) Africa is not what drives Khoisan activism (Jansen, 2016). Rather than demanding recognition as a "superior nation" or non-Khoisan to leave South Africa, it is more common to marshal indigeneity when voicing aspirations for self-determination and larger shares of the country's wealth ("Take a Restorative Approach" 2017; Resoluties van Inheemse Parlement, 2017). As one activist put it to a crowd of sympathisers when discussing the need to embrace indigeneity as a form of self-empowerment: "The Khoisan are dispossessed, the Coloureds are poor" (Verbuyst, 2022: 213). However, opinions also vary widely about how to redress this dispossession. Some lay claim to all South African territory, others focus on reclaiming land of specific historical significance or convincing the government to be prioritised in the land redistribution program (Sato, 2018). Notably, however, virtually all activists make a point of emphasising that all South Africans have a right to belong (see below). Some even reportedly identify as "first indigenous" in order to accommodate the indigeneity of fellow South Africans (De Wet, 2010: 6, 30).

If some do not even regard indigeneity as uniquely Khoisan, what then shapes the boundaries of Khoisan identity? There is no conclusive answer as there are no clear-cut common denominators among Khoisan activists in terms of socio-economic background, race, or creed (see also SAHRC, 2018: 50). This, however, is precisely the point that many Khoisan make when grounding their indigeneity in a non-mutually exclusive combination of having a credible historical experience of being known as 'Coloured' and the willingness to celebrate Khoisan identity and culture. As one activist noted in reference to the Khoisan's marginalisation: "If you have not witnessed a history of oppression, this renders you way down the line in terms of entitlement to indigeneity" (Verbuyst, 2022: 209). Like most others, the same activist reasoned that genetics or descent might make claims to Khoisan identity harder to refute, but they cannot be binding criteria due to centuries of forced assimilation, as well as the fact that not everyone who was labelled 'Coloured' had Khoisan ancestry, and, conversely, that not everyone with Khoisan ancestry was classified as 'Coloured'. It is in the context of traditional leadership disputes or other settings where coveted resources are competed over that disagreements tend to arise over the importance of genetics or other supposed markers of (in)authentic Khoisan identity.[6] Those involved in such conflicts are prone to making contradictory statements and frequently changing their positions. On different occasions, CCHDC spokespersons, for instance, put forward bloodline descent, self-identification, and commitment to Khoisan activism as membership criteria (Verbuyst, 2022: 94–5).

Ethnographic fieldwork tends to lay bare the diversity, dynamism, and contradictions of the social world, and this section has indeed highlighted the variety of emic articulations of indigeneity that exist among Khoisan activists.

In their wide-ranging diversity, emic articulations of indigeneity reflect issues that I earlier raised under the rubrics 'etic-critical' and 'etic-analytical', once again evidencing the conceptual ambiguity at the core of this chapter. As I argue next, distinguishing between articulations of indigeneity is a necessary step to disentangle their use in the current political discourse and offer a refitted theoretical framework based on Zenker's (2021) reading of autochthony and belonging.

Belonging and Conflicting Logics of Autochthony

Zenker's framework is partially based on the notion of 'belonging': feeling and/or being recognised as a member of a group, potentially generating a sense of place, comfort, and safety (see also Chin, 2019). Belonging is context-dependent and relational and can be affected by several factors, such as identity, gender, and nationality (Zenker, 2021: 773). Its meaning is likely subject to a "politics of belonging" as rights and access to resources tend to be at stake (Koot et al., 2019: 347). Being regarded as 'not-belonging' can therefore have devastating consequences. Geschiere (2009) famously spoke of the "perils of belonging" in this regard in reference to various African and European case studies where claims revolving around 'autochthony' – originating from the soil – resulted in violence and discrimination. Crucially, however, Zenker argues that the relationship between autochthony and belonging extends beyond claims of prior occupancy since belonging can derive from a sense of place *and/or* group membership.

For Zenker, autochthony is more accurately understood as referring to two causal logics behind professing a "rightful link between an individual, a territory, and a group" (Zenker, 2021: 779). Both logics affirm the same 'link', but their underlying arguments are diametrically opposed. In the logic of "individualised autochthony", place of birth or residence within a certain territory legitimises individuals' land(ed) rights – i.e., rights to own, cultivate and occupy land *and* rights that flow as a result of 'legitimately' belonging to a certain territory, which in turn links them to a group of fellow rightsholders (Zenker, 2021: 780). Individuals will over time likely share "commonalities of (some) culture and/or (some) descent", but this is not a prerequisite to assert this type of autochthony. "[A]lleged commonality of (some) culture or (some) descent, or both" is, however, essential to the logic of "collectivised autochthony", as these shared traits tie individuals to groups who claim land(ed) rights within a specific territory (Zenker, 2021: 780–1). In many cases, individuals will also be born in the territory in question, but this is not fundamental to the logic of collectivised autochthony.

Both 'logics' assert belonging to place through autochthony, but their apposite premises can have immensely differing implications. This is complicated by the fact that people usually invest in "different circuits of belongings" by engaging "coexisting, overlapping, and contested autochthonies at shifting scales" (Zenker, 2021: 781–4). When looking out for best interests, notions such as place of birth, prior occupancy, group, and culture can easily be conflated. The various articulations of indigeneity I identified in this text indeed similarly compete as conflicting logics of autochthony. Khoisan activists tend to base their emic articulations of indigeneity on the experience of being known as 'Coloured' rather than in being South Africa's undisputed prior occupants. As I have shown, for a minority of activists, these kinds of articulations of indigeneity seem to tap into a wider racialised discourse, giving it the markings of race-based collectivised autochthony claims. This contingent of activists also taps into the same individualised autochthony register that feeds xenophobic assaults on foreign nationals in South Africa at large, i.e., that they have no rights to land because they are not its prior occupants (Solomon, 2019). These Khoisan activists thereby reduce indigeneity to prior occupancy, a form of individualised autochthony. Critics tend to hone in on the same link between indigeneity and prior occupancy, although for different reasons. However, while different interpretations of indigeneity are articulated in practice and there is certainly nothing inherently virtuous about such articulations (Canessa, 2014: 168), indigeneity should ultimately be understood as a collectivised autochthony claim as it pertains first and foremost to group membership. Khoisan activists enjoy land(ed) rights as South African citizens through a logic of individualised autochthony, but this logic apparently leaves their grievances unaddressed. Different articulations of indigeneity and logics of autochthony indeed imply different, if not necessarily contradictory, pathways to decolonisation. In the next section, I put Zenker's framework further to work and argue that indigeneity needs to be *analytically* distinguished from debates over prior occupancy if decolonisation in South Africa is to become more inclusive.

Towards more Inclusive Decolonisation?

The course of decolonisation ultimately needs to be set by all South Africans, not just by the Khoisan or let alone by outsiders such as myself. This contribution is intended to help make more informed assessments and decisions, not to predetermine outcomes or pick sides. I also work with two premises. The first is that the Khoisan's grievances have not been sufficiently addressed in the post-apartheid era. Few would likely disagree that for decolonisation to be as effective as possible it should be as inclusive and comprehensive as possible. I have argued that conceptual ambiguity surrounding indigeneity, which lies at the core of Khoisan activism, constitutes an underexamined impediment. Without an analytical and empirical appraisal of the various articulations of indigeneity, its potential role in decolonisation cannot be properly assessed and the Khoisan's grievances cannot be meaningfully understood or addressed. My second premise is that decolonisation does not necessarily entail rejecting (colonial) concepts and embracing (anti-colonial) alternatives, but can also involve "interpolating" existing discourses (Ashcroft, 2002: 103). Decolonisation should strive towards "pluriversality": the liberating emphasis away from

(Western) abstract universals, and towards a multitude of coexisting ways of being and knowing, in particular those that have been historically neglected and are germane to the local context (Mignolo & Walsh, 2018). A 'pluriversal' approach to indigeneity does not take its meaning for granted or rejects it outright as a foreign or unidimensional concept, but keeps an open mind with regards to how it might be "reverse engineered" and take on various meanings (Niezen, 2003: 221).

In practice, however, South African officials and academics hardly venture beyond 'etic-critical' articulations of indigeneity. Given the country's devastating historical entanglement with racial labels, racialised and exclusionary uses of indigeneity in political discourse justifiably preoccupy the South African government. It might fear radical interpretations of decolonisation which demand complete sovereignty over all ancestral lands (see, e.g., Tuck & Yang, 2012). It does not help the South African debate on decolonisation either that individualised autochthony claims are an increasingly popular way of asserting belonging across the board, particularly in relation to asserting and contesting land rights along racial lines (Kepe & Hall, 2018; Koot et al., 2019; Van-Zyl Hermann & Verbuyst, 2022). Decolonisation based on the prior occupancy of the Khoisan diminishes the historical suffering of countless fellow South Africans. As elsewhere in Africa, the question of indigeneity was therefore likely buried under the weight of the pressing need for decolonisation for the majority (Crawhall, 1999: 324).

Then again, South Africa is certainly more open to debate indigeneity than its neighbours (see, e.g., Hutchison, 2021) and elements within the government and beyond are undoubtedly genuinely committed to making decolonisation more inclusive of the Khoisan. Significantly, the South African Human Rights Commission released a damning report in 2018, which underscored how the Khoisan were still being discriminated against, among others through their continued classification as 'Coloured' or association with hurtful stereotypes (SAHRC, 2018: 9, 17). The Commission recommended a series of measures, including land restitution and officially recognising the Khoisan as a distinct group, but these have not been implemented and the report did not spell out where the Khoisan's marginalisation derives from. Without any material change, a politics of redress lacks credibility (Wells, 2017: 360). The same goes for a process of decolonisation that lacks a clear understanding of how the Khoisan's specific grievances are related to indigeneity; a concept which the report believes could apply to "other African communities" as well (SAHRC, 2018: 8).

While this report is a step in the right direction, it misses the mark in defining indigeneity in a politically correct manner. Such a framing of indigeneity risks contributing to a history of failed policies. Feeling marginalised and misunderstood, the number of Khoisan activists turning to arguments of prior occupancy or race might increase as they feel other articulations of indigeneity are not getting through to policymakers. Khoisan activists certainly disagree over what constitutes indigeneity or what it implies in terms of decolonisation. In the main, however, rather than placing clear-cut (racial) boundaries around their indigeneity, Khoisan activists define it in relation to a lack of belonging and to distinct experiences of marginalisation, which need not be resolved at the expense of fellow South Africans. Claims for recognition and redress can be refracted on a spectrum, ranging from secession and land claims to grievances about affirmative action and enduring systems of racial classification. Crucially, most Khoisan activists do not strive for an exclusive type of decolonisation but explicitly emphasise every South African's right to belong. They might very well tap into different 'circuits of belonging', but collectivised autochthony claims based on race should be separated from those about indigeneity when it comes to designing policy. While it might on occasion act as such in practice, indigeneity is not a stand-in for 'race' as etic-critical perspectives on indigeneity suggest; nor does it relate to the White/Black binary which underpins most discourses on decolonisation. Etic-analytical articulations of indigeneity are currently unpopular but need to be explored more fully. Viewed through this lens, it becomes apparent that Khoisan activists are not primarily concerned with race-based type of decolonisation, but desire another kind centred on marginalisation as a result of settler-colonialism, i.e., past *and* present dispossession and assimilation. In other words: Khoisan activists' emic articulations of indigeneity on balance correspond more to etic-analytical perspectives than etic-critical ones.

To be sure, etic-analytical articulations of indigeneity might also imply a zero-sum game, which is arguably not only undesirable on moral grounds but also complicated by the aforementioned difficulties in describing present-day individuals as descendants of ethnic first-comers; the complicated empirical application of the binary 'settler'/'native' (Veracini, 2010; cf. Wolfe, 2013); and the limits of holding the post-apartheid government accountable for decolonising settler-colonialism (Veracini & Verbuyst, 2020). Taking aim at a different structure of domination (in the South African case, a structure based around 'race'), decolonising settler-colonialism in many ways requires a different approach than decolonising 'colonialism' (Veracini, 2010). However, since both target 'coloniality' – long-standing patterns of power that supersede colonial administrations (see above) – the work of decolonising settler-colonialism can occur alongside other decolonisation schemes in a non-competitive way (cf. Mamdani, 2001). A settler-colonial studies' framework is not primarily concerned with determining prior occupancy, quantifying degrees of suffering, or assessing who merits legal recognition as indigenous people. In fact, somewhat ironically, while revolving around indigeneity in the etic-analytical sense, the Khoisan's grievances could technically be accommodated without referring to the concept itself. As is usually the case with indigenous people (Gausset et al., 2011), Khoisan activists in the main instead place their emphasis on their distinct experiences of marginalisation, which they in turn feel are best captured under the label indigenous. Paradoxically then, the vast majority of Khoisan activists do not reject South African citizenship, which they enjoy by virtue of the logic of individualised autochthony, but seek to render citizenship more meaningful and able to generate belonging by supplementing it with 'indigeneity' as a collectivised autochthony claim. They primarily hold the post-apartheid dispensation accountable, not because they are responsible for past atrocities, but because they are not doing enough to redress these. Without the appropriate understanding of indigeneity in relation to settler-colonialism, the South African race-based decolonisation discourses will continue to misread the Khoisan's distinct marginalisation and unwittingly perpetuate settler-colonial modes of domination.

South African settler-colonialism indeed manifests itself in various largely unacknowledged ways and the Khoisan's input in political discussions on decolonisation pales in comparison to that of academics and government officials. As a result, countless colonial legacies remain undetected and ways of decolonising South African settler-colonialism, which fall beyond the scope of this text, are unexplored. The question of who and why decolonisation is for remains poorly addressed. It is clear that theory, politics, and practice do not exist independently, but as entangled spheres dictating the pace and nature of South Africa's decolonisation. As South Africa continues to reckon with Khoisan activism in the face of conceptual conflation, I have argued that indigeneity can act as a decolonising concept if its diverse uses in political discourse are appraised and additional analytical frameworks are taken on board. In taking indigeneity and discourses on decolonisation out of the realm of individualised autochthony, the post-apartheid dispensation could not only tackle the Khoisan's grievances and marginalisation more effectively, but also deflate the emerging explosive political rhetoric around race-based prior occupancy in its borders. However, it is vital to reemphasise that this decision ultimately falls to the stakeholders that matter most in the debate on decolonisation and indigeneity: South Africans. This is a diverse collective and I only discussed the Khoisan in this text. Further research into how non-Khoisan articulate their indigeneity or how related concepts, such as 'minority', 'African', or 'self-determination' feature in South African political discourse is necessary to refine the analytical toolbox even further and make the decolonisation debate yet more comprehensive.

Notes

1 The term Khoisan (also spelt Khoesan, Khoe-San, or Khoi-San) is contested and offensive to some, yet preferred by others. In this text, 'Khoisan' refers to ethnic groups recognised by the United Nations as the indigenous people of South Africa: the San, Nama, Griqua, Koranna, and Cape Khoi (Stavenhagen, 2005).

2 It should be noted that there is also significant support for the property development among the Khoisan.

3 Scholars disagree over whether Coloured identity was imposed by the colonial government or whether it organically emerged from below (see Adhikari, 2005). In that 'Coloured' refers to a particular shared historical experience, it has certainly become significant to many. It is that historical experience and Coloured identity's connotations of non-indigeneity that are of concern in this text.

4 'Aboriginal' and 'First Nations' are also used among the Khoisan. These designations do not reflect different articulations of indigeneity than those identified in this article, although further research is required.

5 Khoisan activists' articulations of indigeneity of course cannot be reduced to 'politics'. As I have argued at great length elsewhere, their indigeneity also pertains to issues of healing, identity, and culture (Verbuyst, 2022).
6 Tribal affiliations, which regularly cause strive among Khoisan activists, can act as additional layers of self-identification and entitlement.

References

Adhikari, M. (2005). *Not White Enough, Not Black Enough: Racial Identity in the South African Coloured Community*. Athens: Ohio University Press.

Adhikari, M. (2011). *The Anatomy of a South African Genocide: The Extermination of the Cape San Peoples*. Athens: Ohio University Press.

Ashcroft, B. (2002). *Post-Colonial Transformation*. London: Routledge.

Bam, J. (2021). *Ausi Told Me: Why Cape Herstoriographies Matter*. Cape Town: Jacana.

Bam-Hutchison, J. (2016, March). *Lalela: Occupying Knowledge Practices and Processes in Higher Education in South Africa (UCT as Case Study)*. Paper presented at the *Social Anthropology Department at University of Cape Town*, Cape Town.

Barbieri, C., Güldemann, T., Naumann, C., Gerlach, L., Berthold, F., Nakagawa, H., & Pakendorf, B. (2014). Unraveling the Complex Maternal History of Southern African Khoisan Populations. *American Journal of Physical Anthropology, 153*(3), 435–448. https://doi.org/10.1002/ajpa.22441

Besent, M. (2018). Khomani San Distance Themselves from Khoisan Sovereign State Call, *SABC News*, 20 July. Retrieved from https://www.sabcnews.com/sabcnews/khomani-san-distance-selves-khoisan-seceded-call/

Besten, M. P. (2006). *Transformation and Reconstitution of Khoe-San Identities: AAS Le Fleur I, Griqua identities and Post-Apartheid Khoe-San Revivalism (1894–2004)*. Leiden University, Leiden.

Brantlinger, P. (2013). *Dark Vanishings: Discourse on the Extinction of Primitive Races, 1800–1930*. New York: Cornell University Press.

Canessa, A. (2014). Conflict, Claim and Contradiction in the New 'Indigenous' State of Bolivia. *Critique of Anthropology, 34*(2), 153–173. https://doi.org/10.1177/0308275X13519275

Cavanagh, E. (2012). Land Rights that Come With Cut-Off Dates: A Comparative Reflection On Restitution, Aboriginal Title, and Historical Injustice. *South African Journal on Human Rights, 28*(3), 437–457. https://doi.org/10.1080/19962126.2012.11865055

Cavanagh, E. (2013). *Settler Colonialism and Land Rights in South Africa: Possession and Dispossession on the Orange River*. London: Palgrave Macmillan.

Chin, C. (2019). The Concept of Belonging: Critical, Normative and Multicultural. *Ethnicities, 19*(5), 715–739. https://doi.org/10.1177/1468796819827406

Crawhall, N. (1999). Going to a Better Life: Perspectives on the Future of Language in Education for San and Khoe South Africans. *International Journal of Educational Development, 19*(4–5), 323–335. https://doi.org/10.1016/S0738-0593(99)00032-2

De Wet, P. (2010). *South Africa's Unfinished Business. The First Nation Indigenous KhoeSan Peoples*. Saarbrücken: Lambert Academic Publishing.

Department of Land Affairs (DTA). (1997). *White Paper on South African Land Policy April 1997*. Pretoria: Department of Land Affairs.

Erasmus, Z. (2013). Throwing the Genes: A Renewed Biological Imaginary of 'Race', Place and Identification. *Theoria, 60*(136), 38–53.

Francis, D., & Webster, E. (2019). Poverty and Inequality in South Africa: Critical Reflections. *Development Southern Africa*, *36*(6), 788–802. https://doi.org/10.1080/0376835X.2019.1666703

Gausset, Q., Kenrick, J., & Gibb, R. (2011). Indigeneity and Autochthony: A Couple of False Twins? *Social Anthropology*, *19*(2), 135–142. https://doi.org/10.1111/j.1469-8676.2011.00144.x

Geschiere, P. (2009). *The Perils of Belonging: Autochthony, Citizenship, and Exclusion in Africa and Europe*. Chicago: University of Chicago Press.

Huizenga, D. (2018). Articulations of Aboriginal Title, Indigenous Rights, and Living Customary Law in South Africa. *Social & Legal Studies*, *27*(1), 3–24. https://doi.org/10.1177/0964663917710986

Hutchison, Y. (2021). African indigeneity: The Southern African challenge. In G. N. Devy & G. V. Davis (Eds.), *Indigeneity and Nation* (pp. 102–122). London: Routledge. https://doi.org/10.4324/9780429291838

Jacobs, S., & Levenson, Z. (2018). The Limits of Coloured Nationalism, *The M&G Online*, 13 July. Retrieved from https://mg.co.za/article/2018-06-13-00-the-limits-of-coloured-nationalism/

Jansen, L. (2016, November). Untangling Decades Old Myth: Understanding Concept of 'Indigenous Peoples' in Africa. *Eerste Nasie Nuus*, 6–7.

Kamaldien, Y. (2018). "Gatvol Capetonian" Comes under Fierce Attack for Independence Call, *Independent Online*. Retrieved from https://www.iol.co.za/weekend-argus/gatvol-capetonian-comes-under-fierce-attack-for-independence-call-15277270

Kepe, T., & Hall, R. (2018). Land Redistribution in South Africa: Towards Decolonisation or Recolonisation? *Politikon*, *45*(1), 128–137. https://doi.org/10.1080/02589346.2018.1418218

Koot, S., Hitchcock, R., & Gressier, C. (2019). Belonging, Indigeneity, Land and Nature in Southern Africa under Neoliberal Capitalism: An Overview. *Journal of Southern African Studies*, *45*(2), 341–355. https://doi.org/10.1080/03057070.2019.1610243

Kumarakulasingam, N., & Ngcoya, M. (2016). Plant Provocations: Botanical Indigeneity and (De)colonial Imaginations. *Contexto Internacional*, *38*(3), 843–864. https://doi.org/10.1590/s0102-8529.2016380300006

Kuper, A. (2003). The Return of the Native. *Current Anthropology*, *44*(3), 389–402. https://doi.org/10.1086/368120

Langeveldt, W. (2016). *Those Who are First Will Be Last and Those Who Are Last Will Be First! The Story and History of Our People. "Brown People."* s.l.: Korana Kloof Heritage Publishers.

Lehmann, K. (2004). Aboriginal Title, Indigenous Rights and the Right to Culture. *South African Journal on Human Rights*, *20*(1), 86–118.

Makinana, A. (2016). First-Nation Status for Khoisan "Unsustainable", *News24*, 27 November. Retrieved from https://www.news24.com/news24/southafrica/news/first-nation-status-for-khoisan-unsustainable-20161126

Mamdani, M. (2001). Beyond Settler and Native as Political Identities: Overcoming the Political Legacy of Colonialism. *Comparative Studies in Society and History*, *43*(4), 651–664.

Mellet, P. T. (2010). *Lenses on Cape Identities: Exploring Roots in South Africa*. Cape Town: Dibanisa Publishing.

Mignolo, W., & Walsh, C. (2018). *On Decoloniality. Concepts. Praxis. Analysis*. Durham and London: Duke University Press.

Ndlovu-Gatsheni, S. J. (2009). Africa for Africans or Africa for "Natives" Only? "New Nationalism" and Nativism in Zimbabwe and South Africa. *Africa Spectrum, 44*(1), 61–78. https://doi.org/10.1177/000203970904400105

Ndlovu-Gatsheni, S. J. (2013). Why Decoloniality in the 21st Century. *The Thinker, 48*(10), 5–9.

Niezen, R. (2003). *The Origins of Indigenism: Human Rights and the Politics of Identity*. Oakland: University of California Press.

Niezen, R. (2009). *The Rediscovered Self: Indigenous Identity and Cultural Justice*. Montreal: McGill-Queen's Press.

Observatory Civic Association et al. v Trustees for the time being of Liesbeek Leisure Properties Trust et al. [2022] High Court of South Africa, Western Cape Division, Cape Town, Case No: 12994/21.

Online Readers React. (2013, December). *Eerste Nasie Nuus*, p. 7.

Pelican, M. (2009). Complexities of Indigeneity and Autochthony: An African Example. *American Ethnologist, 36*(1), 52–65. https://doi.org/10.1111/j.1548-1425.2008.01109.x

Radcliffe, S. A. (2018). Geography and Indigeneity II: Critical Geographies of Indigenous Bodily Politics. *Progress in Human Geography, 42*(3), 436–445. https://doi.org/10.1177/0309132517691631

Ramaphosa, C. (2019). Address by President Cyril Ramaphosa at the Annual Opening of the National House of Traditional Leaders, Parliament, Cape Town. 19 February. Retrieved from https://www.gov.za/speeches/president-cyril-ramaphosa-opening-national-house-traditional-leaders-19-feb-2019-0000

Rassool, C. (2019). The Politics of Nonracialism in South Africa. *Public Culture, 31*(2), 343–371. https://doi.org/10.1215/08992363-7286861

Resoluties van Inheemse Parlement. (2017, August). *Eerste Nasie Nuus*, p. 5.

Rose, D. B. (1996). Land Rights and Deep Colonising: The Erasure of Women. *Aboriginal Law Bulletin, 3*(85), 6–13.

Sato, C. (2018). Khoisan Revivalism and Land Question in Post-Apartheid South Africa. In F. Brandt & G. Mkodzongi (Eds.), *Land Reform Revisited. Democracy, State Making and Agrarian Transformation in Post-Apartheid South Africa* (pp. 134–155). Leiden and Boston: Brill.

Settler, F. (2010). Indigenous Authorities and the Post-Colonial State: The Domestication of Indigeneity and African Nationalism in South Africa. *Social Dynamics, 36*(1), 52–64. https://doi.org/10.1080/02533950903561239

Solomon, R. (2019). Xenophobic violence and the Ambivalence of Citizenship in Post-Apartheid South Africa. *Citizenship Studies, 23*(2), 156–171. https://doi.org/10.1080/13621025.2019.1584158

South African Human Rights Commission (SAHRC). (1999). *Research Project on the Rights of Indigenous Peoples*. Pretoria: South African Human Rights Commission.

South African Human Rights Commission (SAHRC). (2018). *Report of the South African Human Rights Commission: National Hearing Relating to the Human Rights Situation of the Khoi-San in South Africa*. Pretoria: South African Human Rights Commission.

Stavenhagen, R. (2005). *Report of the Special Rapporteur on the Situation of Human Rights and Fundamental Freedoms of Indigenous People – Mission to South Africa*. Geneva: United Nations.

Stop calling us 'coloured' and denying us our diverse African identities. (2018). *Mail & Guardian*, 2 August. Retrieved from https://mg.co.za/article/2018-08-02-stop-calling-us-coloured-and-denying-us-our-diverse-african-identities/

Take a Restorative Approach. (2017, March). *Eerste Nasie Nuus*, p. 13.

Tuck, E., & Yang, K. W. (2012). Decolonisation is Not a Metaphor. *Decolonisation: Indigeneity, Education & Society*, *1*(1), 1–40.

Van-Zyl Hermann, D., & Verbuyst, R. (2022). 'The Real History of the Country'? Expropriation without Compensation and Competing Master Narratives about Land (Dis)Possession in South Africa. *Journal of Southern African Studies*, *48*(5), 825–844. https://doi.org/10.1080/03057070.2022.2127565

Veracini, L. (2010). *Settler Colonialism: A Theoretical Overview*. London: Palgrave Macmillan.

Veracini, L. (2011). Isopolitics, Deep Colonizing, Settler Colonialism. *Interventions*, *13*(2), 171–189. https://doi.org/10.1080/1369801X.2011.573215

Veracini, L., & Verbuyst, R. (2020). South Africa's Settler-Colonial Present: Khoisan Revivalism and the Question of Indigeneity. *Social Dynamics*, *46*(2), 259–276. https://doi.org/10.1080/02533952.2020.1805883

Verbuyst, R. (2016). Claiming Cape Town: Towards a Symbolic Interpretation of Khoisan Activism and Land Claims. *Anthropology Southern Africa*, *39*(2), 83–96. https://doi.org/10.1080/23323256.2016.1171159

Verbuyst, R. (2022). *Khoisan Consciousness: An Ethnography of Emic Histories and Indigenous Revivalism in Post-Apartheid Cape Town*. Leiden and Boston: Brill.

Wells, J. C. (2017). When the Past Transforms: A Case Study from a Western Cape Wine Farm. *South African Historical Journal*, *69*(3), 345–360. https://doi.org/10.1080/02582473.2017.1332084

Wolfe, P. (2006). Settler Colonialism and the Elimination of the Native. *Journal of Genocide Research*, *8*(4), 387–409. https://doi.org/10.1080/14623520601056240

Wolfe, P. (2013). Recuperating Binarism: A Heretical Introduction. *Settler Colonial Studies*, *3*(3–4), 257–279. https://doi.org/10.1080/2201473X.2013.830587

Zenker, O. (2021). Politics of Belonging. In M. C. Foblets, et al. (Eds.), *The Oxford Handbook of Law and Anthropology* (pp. 772–791). Oxford: Oxford University Press. https://doi.org/10.1093/oxfordhb/9780198840534.013.45

Zuma, J. (2001). Address by Deputy President Zuma to the opening ceremony of the National Khoisan Consultative Conference.

Zuma, J. (2013). State of the Nation Address, 14 February. Retrieved from https://www.politicsweb.co.za/documents/jacob-zumas-state-of-the-nation-address-2013

Afterword

Ritu Vij

Decolonising Political Concepts joins a growing inter-disciplinary line of enquiry into the constitutive role of colonialism and empire in the making of the modern world. The onto-political concepts that compose the architecture of political modernity – sovereignty, democracy, freedom, the notion of the subject, and agency, among others – come under scrutiny here for the occlusion of the hidden transcripts of imperial and colonial violence that underpin modern liberal and heterodox political imaginaries. The fictions of a shared political horizon of autonomy, freedom, and justice, belied by the history and after-lives of slavery, capitalist extractive logics, and the fractal divisions of who counts or does not count as properly human are brought into sharp visibility in this collection. In alignment with the desire of the decolonial collective to call an end to the imperialism of categories that enable the dominance of "Western reason" and the denial of coevalness with other (i.e., non-Western) geographies of reason, this volume aims to decolonise political concepts from Euro-modernity's enclosure of reason *tout court*. Defying the high priests of the Western canon of political thought mobilises a shared desire to recover modes of thinking in the hinterlands of modernity to acknowledge and build on subaltern political worldings. The recovery of suppressed knowledges holds out the promise of subverting one-world thinking, countering Hobbesian narratives of the civilising effects of war and state-making and Lockean claims of private property as the necessary predicate of sovereign subjectivity (both, incidentally, shareholders in colonial trade companies). In a neat reversal of stadial narratives of progress in which the direction of travel is only ever from the West to the East, *Decolonising Political Concepts* engages – and extends – Latin American Decolonial Theory to consider how non-European geo-epistemologies can replenish political thinking in the contemporary conjuncture. In the brief reflections that follow, I call attention to some of the more generative lines of thinking outlined in the book.

The chief strengths of this book are (1) its fidelity to the ambition that constellates the body of work produced by the Latin American theorists widely regarded as progenitors of Decolonial Theory (DT) (Enrique Dussel, Aníbal Quijano, Walter Mignolo, Nelson Maldonado-Torres, Ramon Grosfoguel, Boaventura de Sousa Santos). As outlined in the Introduction, Modernity,

DOI: 10.4324/9781003293460-13

Coloniality, and Decoloniality work in tandem in very specific ways. Because modernity and coloniality are constitutively entangled and therefore inseparable, all roads beyond DT's critique of modernity and coloniality point to the pre-modern and pre-colonial as the exclusive register of decoloniality proper. (2) Thinking both with and against the grain of Decolonial Theory, the project undertaken in this collection questions the Western canon for its elision of the inseparability of modernity from colonialism but also presses against disciplining proclivities within the Decolonial canon as well. (3) The book's singular achievement, however, is its extension of decolonial thought beyond its Latin American provenance. Deploying DT in diverse contexts, the empirically rich discussions of political concepts and praxis in South Africa, the Southern Mediterranean, the Middle East, and South-Asia bring nuance and clarity to ongoing debates around contested political forms, including the category of the Indigenous, central to Decolonial Theory.

Foregrounding the colonial unconscious (Traverso, 2016: 174) that underpins political concepts and ordering practices, the collection attempts to show "how predominant political concepts in Western political thought are beset with colonial remnants in their construction, formulation, and deployment". This, the editors suggest, leads to "naturalising" political life such that the depredations that ensue from the practices of exclusion/inclusion that are integral to the containment of political imaginaries in nation-states, capitalist markets, and individuated subjectivities, fall outside the purview of political life. Contributors to this volume take up Decolonial Theory's critique of Western metaphysics and the Eurocentric epistemologies that work to produce and rationalise systemic injustice to show how modern political concepts as de-politicised modes of domesticating political difference foreclose vernacular political idioms of thought. Thus, Cecilia Cienfuegos Martínez, drawing on Hortense Spillers and Maria Lugones, examines the coloniality of gender to critique conventional de-politicised general paradigms of sexual difference. Calling attention to sexual violence as pre-eminently political, given the "locality of violence", Cienfuegos Martínez's argument helps re-locate the problem of sexual violence from its normalised registers of inter-personal relations, or as a technology of war in conflict zones. Similarly, Laurencia Saénz's critical re-formulation of "white ignorance" as principally conative not cognitive mobilises DT's desire to uncover the "colonial remnants" in the categories of thought that structure contemporary life. It also, however, presses on DT's concern with epistemology and sounds a note of caution regarding the redemptive potential of "epistemological disobedience" alone. Engaging emotions and affects, not just cognition, Sáenz suggests, is necessary to undoing white ignorance. Finally, to decolonise the notion of agency, Henrike Kohpeiß and Marie Wurth develop a critique of Hannah Arendt's notion of agency to uncover the impossibility of action and agency for enslaved populations that ensues from the coloniality of power. Drawing on notions of performative action and practice in the work of Saidiya Hartman and Fred Moten, in particular, the turn to Black Thought enriches the volume's investment in exposing the "colonial load" that Western political thought carries.

Resisting DT's wholly negative gesture of critique and the injunction to turn exclusively to non-Western epistemologies to anchor political struggle and transformation, contributors to this volume excavate modern political practices to call attention to how they can be effectively deployed to effect emancipatory decolonial outcomes. Exemplary in this regard is the preeminent political concept of the state which remains perilously tied to the fate of postcolonial societies, both rejected and affirmed. The post-colonial state is both and at once the only institutional source of relief for subaltern populations in the non-West, but also, as critics of the state-form have long argued, the apotheosis of what Foucault (1990) termed the "modern episteme". Any critical approach to the problem of the political, especially in the post-colony needs then to grapple with the state as *pharmakon*. Unlike canonical Decolonial Theory which elects to dispense with *all* modern political concepts, including the state, contributors to this volume traverse a more generative path. "Political concepts are not neutral or innocuous but explicit tools of power and efficient vehicles for establishing or changing relations of domination", the editors note. "As historical constructs, they are part of the colonial legacies that still permeate our contemporary world … Yet, at the same time, they may be articulated and put to work in ways that may trouble" the "colonial load" they carry. Putting to work extant political concepts, Saxena and Chitkara's incisive critique of the hegemonic understanding of state sovereignty as a singularity uncovers the plurality and co-constitution of rule-making authority in the Indian context. Arguing that sovereignty, "like any other power is characterised by multiplicity", Saxena and Chitkara find affordances in Indian legal jurisprudence to advance Adivasi claim-making. Eschewing the binary framing of state and Indigenous sovereignty characteristic of DT, this contribution outlines an alternative approach that sheds light on the horizontal relation between sovereignty and Indigenous autonomy. Similarly, Shahin Nasiri develops a counter-history of nation-states to suggest that "refugeehood", conventionally understood as inhabiting an anomalous juridico-political space in relation to static understandings of state-based citizenship is better seen as a site for developing alternative notions of (un) freedom and (non)subjectivity. Through flight and multiple border crossings, refugees resist the exclusionary structure of unfreedom – what Fanon might call non-being – to enact their freedom and subjectivity. The state-based political order of the modern world does not present an unsurpassable horizon here: within its crevices and fault lines, practices of resistance signify epistemic and subjective creativity, not simply erasure.

Contributors to this volume thus join a trans-disciplinary scholarship that refuses binary distinctions between the colonial-modern and the Decolonial in a shared effort to apprehend the uneven and mobile workings of the (colonial) "modern episteme" in diverse contexts. Undoing the epistemic ravages of modernist thought and the world fashioned in its image entails not simply repudiation and disavowal at a rhetorical or theoretical level, but the far more difficult task of understanding how people and places "live with concepts". In this regard, the collection echoes the noted anthropologist Veena Das' long-standing commitment to a philosophically inspired ethnographic recuperation of concepts that "emerge out of the engagement with practices of everyday life" (2020: xiii). For Das, as for the authors in this volume, this helps track political concepts, including the everyday state in all its quotidian manifestations in ordinary life. Concepts, Das notes, are not "magic words" that "open up a region of thought and illuminate empirical observations as with the touch of a button" (2020: xiii). Nor are they, for purposes of the project here, impositions that colonise the totality of the everyday. Rather, they are rethought in the vernacular idioms that prevail at different geo-sites, owned, accessed, and re-shaped in ways that are not – and cannot – be predetermined either by their abstracted form or through vernacular idioms elsewhere.

Against Audre Lorde's widely noted scepticism about whether the master's tools can be effectively used against the master, the book mobilises also the sentiment expressed in Gayatri Spivak's provocation to use Enlightenment thought "from below" which opens up the possibility of critiquing Western political thought "from within, to turn it away from itself" (Spivak, 1999: 49). Whereas for Spivak the challenge is to explore if the "magisterial texts" of the progenitors of the dominant political concepts that have shaped modernity – Kant, Hegel, and Marx, most notably – "can now be made our servants" (Spivak, 1999: 6–7), for the editors and contributors to this volume, the more pressing task is the problem of political transformation and how to think about the conditions of possibility, both conceptual and practical, necessary for developing and materialising emancipatory political imaginaries.

Foregrounding the need to "work through, re-think, or even overcome" dominant political thought in order to "transform it into an (a)venue for postcolonial and decolonial struggles", the collection attempts to shed light on the fraught and contested entanglements with modern political forms, rather than dispense with them altogether as DT does. It calls attention to "what is still a blind spot in decolonial theory, while simultaneously introducing a decolonial perspective in multiple discourses and analyses across the social sciences and humanities". What sets this collection apart from standard decolonial endeavours is the recognition that decoloniality entails not just the undoing of the "epistemicide" engendered by colonial modernity, but also the intractable problem of grappling with the enduring enchantment of modernity in and for the very populations in whose name Decolonial Theory speaks.

The ontologisation of modern political concepts, the (re)making of many worlds into a world comprised of sovereign nation-states, capitalist markets, and sovereign subjects, has generated the unsettling paradox of formally decolonised societies' evident embrace of the regulative ideals, political concepts, and crucially, political forms, incubated in colonial modernity. Even as the depredations of colonial modernity in social formations in the Global South are hidden in plain sight, this attachment remains unshakeable. Post-colonial aspirations and roadmaps, as Arjun Appadurai once put it, unfold within a dialectic of desire for and resistance against "modernity" in all its symbolic and material complexity. These aspirations, however, as this volume correctly intuits, are not only the purview of post-colonial elites but also part of the horizon of subaltern aspirations. In their refusal to dismiss the non-West's fraught desire for and against modernity, including its political modalities, as false consciousness or as merely mimetic, contributors respond to the challenge of exploring the entanglements, antinomies, and contestations between modern political forms and praxis and the vernacular idioms through which these modalities are engaged or resisted.

Importantly, the volume is also attentive to the thorny questions of translation, context, and history in its attempt to mobilise and re-work a decolonial analytic. If political concepts forged in the development of a colonial modernity "travel" only by the complete erasure of local forms of political life and ways of knowing as Decolonial Theory is wont to claim, attempts to generalise from the Latin American provenance of Decolonial Theory can be legitimately seen, as one scholar notes, as "limited in its understanding of the problem of colonialism and should therefore not be universalised as the way to theorise the problem of colonialism" (Pillay, 2021: 391). For Decolonial Theorists, the problem of epistemological erasure and its undoing by a turn to Indigenist cosmologies and movements is key to their project. Writing from the African context of apartheid rather than the settler-colonialism of Latin America, however, Pillay's sympathetic critical engagement (and it is but one of many) with DT follows a parallel approach to the one sketched here: to situate and historicise decoloniality in its varied perturbations. The key takeaway from Pillay's critique is the injunction to think conjuncturally: critical political interventions in thought and action are always situated in time and place. The return to a pre-colonial and pre-capitalist past invoked by DT as a solution to the enduring violence produced by the colonial encounter is set aside here as several contributors pursue a fine-grained approach to the politics of decoloniality.

Illustrating this attention to context, Rafael Verbuyst's chapter explores the politics of indigeneity in the South African context. Highlighting the pitfalls of an uncritical embrace of Indigeneity as a category of liberation, Verbuyst carefully charts the fraught politics of Indigeneity in South Africa that suborn the claims of the Khoisan, an unacknowledged ethnic group, to a wider (decolonial) claim that *all* Africans are Indigenous now. This "colonial equivocation" enables an exclusionary articulation of indigenous politics in the context of South Africa anchored, paradoxically, in a wider politics tethered to claims about the decolonisation of the majority. Likewise, "challenging the univocity of concepts, their history, and their uses", Laura Galián's discussion of the translation of anarchist thought into Arabic vernaculars in Egypt, Morocco, and Lebanon shows how decolonised modes of resistance emerge from a linguist praxis of translation. The extension of anarchist thought in translation re-locates national aspirations for liberation in a global context of justice, unsettling the borders that separate the geographies and histories of colonial *and* decolonial reason. And finally, Karim Barakat turns to historical analysis and the social conditions under which political concepts take shape to address the charge of relativism often levied against DT. While some might baulk at Barakat's call to develop "an objective conception of history" dependent on a method or set of criteria that would enable adjudication between different historical views, the chapter succeeds in sounding a cautionary note about the universalising impulse endemic to political philosophy and political theory. Thinking decolonially, for Barakat, warrants a turn both to and away from particular histories.

Unlike Decolonial Theorists for whom the project of undoing epistemological erasure of non-European, specifically Indigenous, ways of thinking demand a clean rupture from colonial modernity by an embrace of the virtues of a pre-capitalist or pre-colonial world, this collection eschews a binary framing of the colonial/decolonial. Rather, it takes the relational history of coloniser and colonised seriously in its attempt to explore the politics of decolonising the historical present. Entanglements between coloniser and colonised, the master and the slave, the oppressor and the oppressed *on both sides of the line* (between Europe and the non-European) offer a decolonial analytic potentially far more generative of the solidarity, humanity, and conviviality that DT seeks to resurrect but paradoxically forecloses by de-linking and un-coupling the previously colonised (Latin America in this case) from the Euro-modern world. For thinkers like Achille Mbembe (2021), for instance, people on the margins of all societies – the global subalterns, one might say – inhabit a similar and shared space of exclusion from the vectors of society. In this space at the extremes of society, those denied their humanity can create zones of *créolité*, spaces of dialogue, and creativity to forge trans-local, trans-national political imaginaries, and bonds of sociality that offer road maps for World-making after empire. Within these *créolité* spaces, new forms of folk politics and political imaginings interact with extant modes of political association (including the State in its inordinate and everyday register), to enable new political concepts and practices to emerge not only locally, but, more urgently, internationally. Harking back to the Bandung moment of 1955 in which leaders like Nasser, Sukarno, Nehru, and Nkrumah attempted to forge a new way of being in the world from the erstwhile space of what Frantz Fanon referred to as nonbeing, ongoing historical work devoted to recovering alternative universalisms embedded in black citizenship (as envisioned in the Haitian Revolution and a trans-national anti-colonial praxis, for instance) aligns with theoretical attempts to develop political concepts and practices enabling of an emancipatory politics.

Echoing Habermas' declaration of modernity being an unfinished project, Maldonado-Torres (2011), one of the principal theorists of DT has also announced that decoloniality is an "unfinished project". If the former has been critically received as a weak defence of the dark underbelly of modernity, specifically its imbrication with colonial and racial violence, the latter can be seen as a pre-emptive defence of the aporetic claims of Decolonial Theory (i.e., in its Modernity/Coloniality/Decoloniality iteration). If decoloniality proper warrants the return to a pristine, uncontaminated pre-modern and pre-colonial past, this return is condemned to a permanent deferral by continued attachments to the very shape of the world DT hopes to unravel. Unlike Olufemi Taiwo's (2022) call to dispense with the "ideology of decolonisation" altogether, on the grounds that it disavows African agency and provides an alibi for political ineptitude by African political elites, this collection's sustained attunement to the enchantments and disenchantments with modernity in the post-colony render the radical decolonisation on offer by DT at best utopian, at worst naïve. In a world in which nation-states, markets, and the rights-bearing subject constitute the regulative horizon of the world, the spirit of decolonising the "modern episteme", and transforming nonbeing into being requires more than the articulation of pre-modern genealogies of thought. For bringing readers to the threshold of thinking decoloniality contrapuntally, this volume deserves notice.

In closing, it is worth noting that the question of whether decolonising political concepts can deliver an emancipatory horizon is inseparable from how a reflexive decolonial project can help foster new political forms through contestation in socio-political struggles.

References

Das, V. (2020). *Textures of the Ordinary: Doing Anthropology after Wittgenstein*. New York: Fordham University Press.

Foucault, M. (1990). *Les mots et les choses*. Paris: Gallimard.

Maldonado-Torres, N. (2011). Thinking through the Decolonial Turn: Post-Continental Interventions in Theory, Philosophy, and Critique--An Introduction. *Transmodernity: Journal of Peripheral Cultural Production of the Luso-Hispanic World, 1*(2), 1–15.

Mbembe, A. (2021). *Out of the Dark Night*. New York: Columbia University Press.

Pillay, S. (2021). The Problem of Colonialism: Assimilation, Difference, and Decolonial Theory in Africa. *Critical Times, 4*(3), 389–416.

Spivak, G. (1999). *A Critique of Postcolonial Reason: Toward a History of the Vanishing Present*. Cambridge: Harvard University Press.

Taiwo, O. (2022). *Against Decolonisation: Taking African Agency Seriously*. London: C. Hurst and Company.

Traverso, E. (2016). *Left-Wing Melancholia: Marxism, History, and Memory*. New York: Columbia University Press.

Index

Pages followed by "n" refer to notes.

For Product Safety Concerns and Information please contact our EU
representative GPSR@taylorandfrancis.com
Taylor & Francis Verlag GmbH, Kaufingerstraße 24, 80331 München, Germany

www.ingramcontent.com/pod-product-compliance
Ingram Content Group UK Ltd.
Pitfield, Milton Keynes, MK11 3LW, UK
UKHW020003200926
475702UK00015B/2452

* 9 7 8 1 0 3 2 2 7 5 9 7 0 *